VARIABLE SPEED PUMPING

A GUIDE TO SUCCESSFUL APPLICATIONS

Variable Speed Pumping

A Guide to Successful Applications

Europump and
Hydraulic Institute

ELSEVIER

UK Elsevier Ltd, The Boulevard, Langford Lane, Kidlington, Oxford OX5 1GB, UK
USA Elsevier Inc, 360 Park Avenue South, New York, NY 10010-1710, USA
JAPAN Elsevier Japan, Tsunashima Building Annex, 3-20-12 Yushima, Bunkyo-ku, Tokyo 113, Japan

British Library Cataloguing in Publication Data
A CIP Catalogue record for this book is available from the British Library

ISBN 1-85617-449-2

Published by
Elsevier Advanced Technology
The Boulevard, Langford Lane, Kidlington, Oxford OX5 1GB, UK
Tel: +44(0) 1865 843000
Fax: +44(0) 1865 843971

Typeset by Land & Unwin
Printed and bound in Great Britain by MPG Books Ltd, Bodmin, Cornwall.

Table of contents

Chapter • One
Introduction

Chapter • Two
Pumping system hydraulic characteristics

Chapter • Three
System and process requirements

Chapter • Four
Pumps

Chapter • Five
Concepts for estimating pumping energy costs

Chapter • Six
Motors

Chapter • Seven
Variable speed drives

Chapter • Eight
Control principles for variable speed pumping

Chapter • Nine
Selection process – new systems

Chapter • Ten
Selection process – retrofitting to existing equipment

Chapter • Eleven
Benefits, drawbacks and operational issues

Chapter • Twelve
Financial justification

Chapter • Thirteen
Case studies

Appendix A1
Electric motors

Appendix A2
Frequency converters

Appendix A3
Legal obligations and harmonic suppression

Appendix A4
Frequently asked questions

Appendix A5
Effects of pump speed and impeller diameter on magnetically driven pumps

Appendix A6
Efficiency of variable speed drive systems

Appendix A7
Non VFD drives

Appendix A8
Glossary

CHAPTER ONE

Introduction

1.1 Objectives of the guide

Variable speed drives in pump systems are now a mature technology, which can generate large benefits to the user in cost savings and reliability improvements, in the right applications. However it is apparent that opportunities for lower life cycle costs are being missed by continuing to use traditional control methods. Conversely there are examples of ineffective systems and wasted expenditure, caused by wrongly applying variable speed drives.

A study was carried out to find what information was available on this subject, and it became apparent that very little was specific to variable speed driven pumps. Another observation was that the few articles found, were focused on either electrical or hydraulic issues, and the interaction of the component parts of the complete system was not described.

The conclusion was that there was a real need for a comprehensive guide on variable speed pumps, motors and drives.

A further major catalyst in forming the international group of mechanical and electrical experts to create this guide, was the need for energy conservation. This not only saves cost to users, but also has environmental benefits.

From studies carried out by the European Commission, it was shown that pumping systems account for nearly 22% of the world's electric motor energy demand. In certain industrial plants over 50% of electrical energy used by motors can be for pumping systems. Figure 1.1 shows a typical split of energy usage of motor driven equipment on an industrial plant. Clearly there is scope for major savings by operating pumps more effectively and efficiently.

Figure 1.2 shows the breakdown of energy consumption by the two main pump categories. Rotodynamic pump types account for 73% and positive displacement pumps 27%, of total energy used by pumps. The

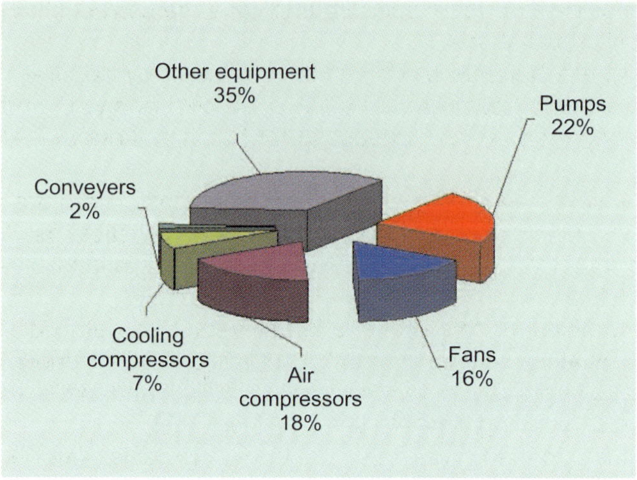

Figure 1.1: Share of motor electricity consumption[1]

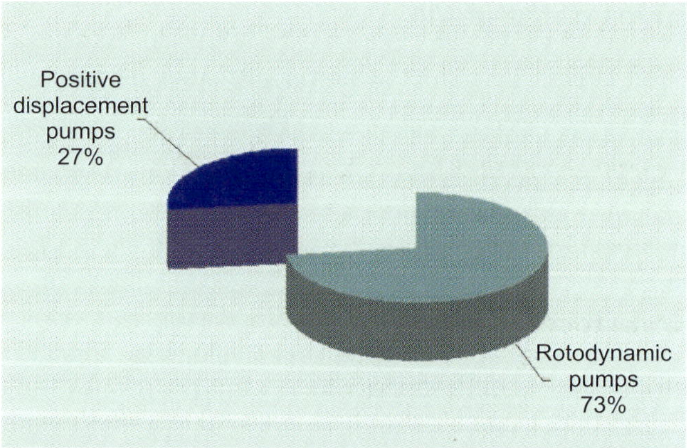

Figure 1.2: Energy consumption by pump type

two types have quite different characteristics in response to varying operating speed, and both are described fully in this guide.

In constructing this document, the group has tried to offer hydraulic, mechanical and electrical assistance to the reader who wishes to evaluate variable speed driven pumps.

This guide includes sections on pumps, systems, motors, variable speed drives and control methods, which start with the basic principles, but progress to more advanced and detailed concepts. At all times the importance of understanding the process requirements and system design is emphasised, and the interrelation of the various elements is explained.

When the requirements of the pump and system are defined, the use of our flow charts will enable the reader to work through a logical selection

[1] *European Commission SAVE Study on Pumps*

process to decide if variable speed pumping is the correct choice for either new or retrofit installations.

The long term benefits are sometimes missed by apparent high initial costs, but we show how to make a life cycle cost analysis, and even how the investment need be no higher than for traditional pump system control methods.

It is essential to avoid inappropriate applications. Therefore we explain the limitations, drawbacks and operational characteristics, as well as highlighting the importance of reviewing potential installations on a case-by-case basis.

The appendices of this guide offer a wealth of detailed information, including legislative issues. A number of case studies show achievable cost savings, improved pump system reliability and process efficiency improvements, by using variable speed drives.

This guide concentrates on electronic variable frequency drives because they have the greatest benefits, including wide ranging applications, versatile control, energy saving and ease of retrofitting.

In conclusion we hope that this guide encourages the use of variable speed pumping, in appropriate applications, leading to cost savings from both reduced energy consumption and increased pump system reliability.

John Bower
Manager of Engineering and Technology
Flowserve UK
(Chairman of the VSD Group)

1.2 Acknowledgements

The International working group was formed from members of the pump, motor and drives industries and it is to these individuals we are all indebted for this guide.

John Bower
Manager of Engineering and Technology
Flowserve UK
(Chairman of the VSD Group)

Steve Schofield
Director, Technical Services
BPMA
(Secretary of the VSD Group)

The Chairman and the Secretary of the working group wish to thank all the team members for their contributions.

In particular a special mention should be given to the main working group members:–

Geoff Brown – ABB Ltd
Gunnar Hovstadius – Consultant
Lez Warren – Cat Pumps
Ian James – HMD Sealless Pumps
Steve Allen – Johnson Pumps (Viking)
Zbigniew Czarnota – ITT Flygt
Aldo Janigro – Assopompe
Hansjurgen Kech – Wilo AG
John Thorne – ITT Industries
Manfred Sacher – Vogel Pumpen

Assistance with the guide was also received from:–

Gregory Romanyshyn – Hydraulic Institute
Jean-Francois Lapray – Alstom
Charles Cappellino – ITT Goulds
Michel Laroche – Wilo AG
Bill Beekman – Floway
Jim Mahon – Lovejoy Inc
Friedrich Klutsch – VDMA
Stefan Abelin – ITT Flygt
Vestal Tutterow – Alliance to Save Energy
Julian Atchia – SJE-Rhombus
Mick Cropper – Sulzer Pumps
Tony Garcia – ITT Bell & Gossett
Manfred Belsdorf – Sterling Fluid Systems
Tim Rahill – Rockwell Automation
David McKinstry – Imo Pump
Aimee Mckane – LBNL
Steve Cummins – SR Drives

The following independent reviewers kindly agreed to give comments and opinions on the guide during its final stages:-

Professor Girobone – Genoa University – Italy
Paul Barnard – Deepdale Consulting Ltd – UK
Roger Lawrence – RGL Solutions – USA
Dennis Rusnak – Flowserve – USA
Maurice Yates – Advanced Energy Monitoring Systems – UK
Dr Hugh Falkner – Future Energy Solutions – UK
Professor Almeida – Coimbra University – Portugal

CHAPTER TWO

Pumping system hydraulic characteristics

2.1 System characteristics

In a pumping system, the objective, in most cases, is either to transfer a liquid from a source to a required destination, e.g. filling a high level reservoir, or to circulate liquid around a system, e.g. as a means of heat transfer.

A pressure is needed to make the liquid flow at the required rate and this must overcome losses in the system. Losses are of two types: static and friction head.

Static head is simply the difference in height of the supply and destination reservoirs, as in Figure 2.1. In this illustration, flow velocity in

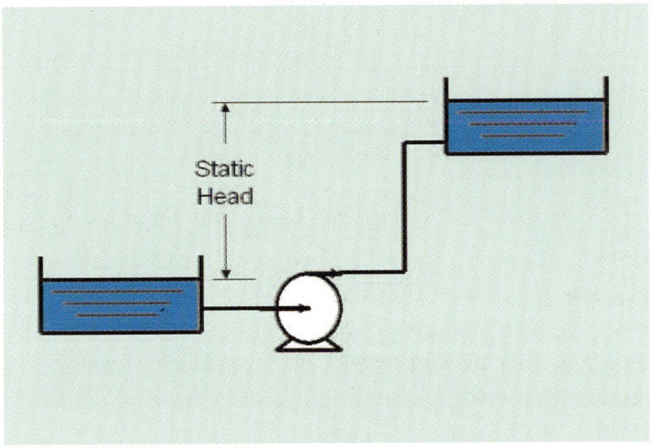

Figure 2.1: Static head

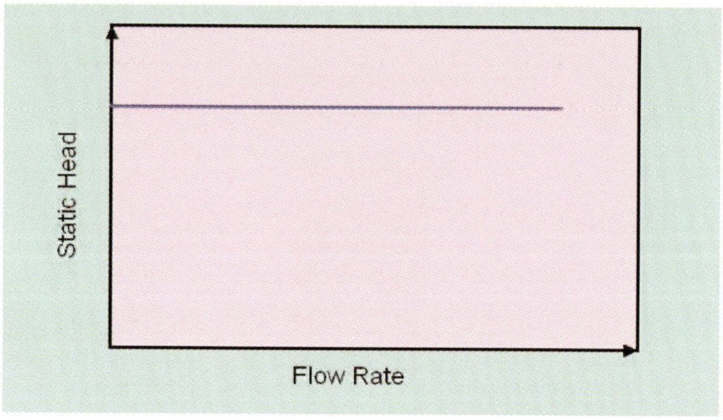

Figure 2.2: Static head vs. flow rate

the pipe is assumed to be very small. Another example of a system with only static head is pumping into a pressurized vessel with short pipe runs. Static head is independent of flow rate and graphically would be shown as in Figure 2.2.

Friction head (sometimes called dynamic head loss) is the friction loss, on the liquid being moved, in pipes, valves and equipment in the system. These losses are proportional to the square of the flow rate. A closed loop circulating system without a surface open to atmospheric pressure, would exhibit only friction losses and would have a system head loss vs. flow characteristic curve as Figure 2.3. This is known as a system curve.

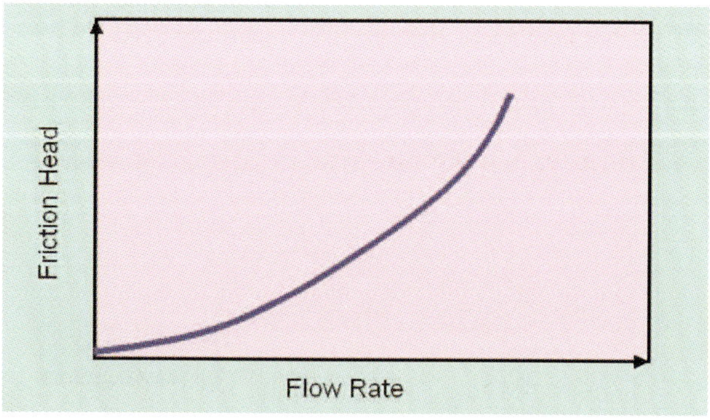

Figure 2.3: Friction head vs. flow rate

2.2 System curves

Most systems have a combination of static and friction head and the system curves for two cases are shown in Figures 2.4 and 2.5. The ratio of static to friction head over the operating range influences the benefits achievable from variable speed drives (see Section 4.2.2.2)

Static head is a characteristic of the specific installation and reducing

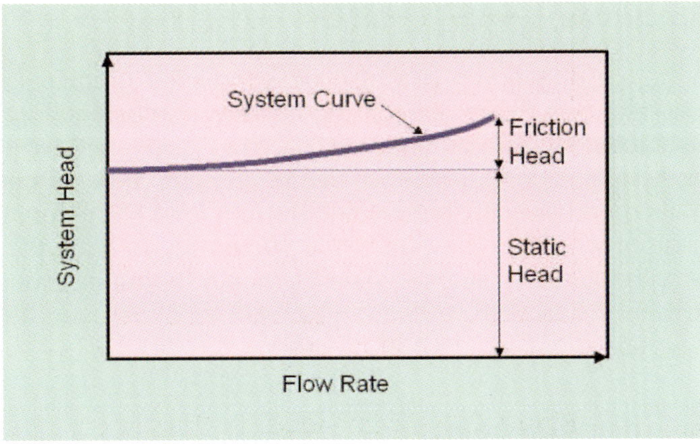

Figure 2.4: System with high static head

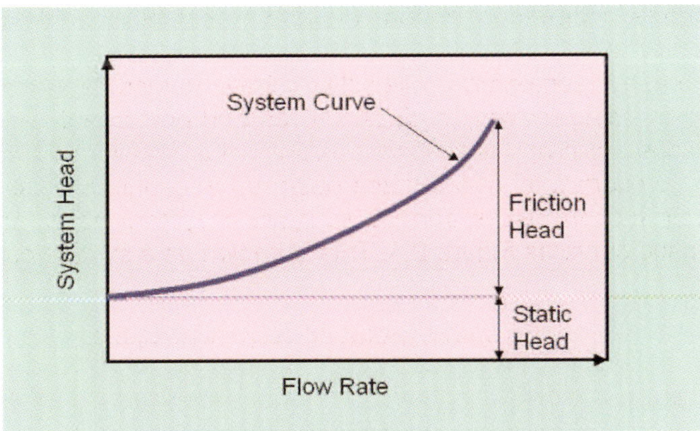

Figure 2.5: System with low static head

this head, where this is possible, generally helps both the cost of the installation and the cost of pumping the liquid. Friction head losses must be minimized to reduce pumping cost, but after eliminating unnecessary pipe fittings and length, further reduction in friction head will require larger diameter pipes, which adds to installation cost.

2.3 Pump curves

The performance of a pump can also be expressed graphically as head against flow rate. See Figure 2.6 for rotodynamic pumps and Figure 2.7 for positive displacement (PD) pumps.

The rotodynamic pump, (usually a centrifugal pump) has a curve where the head falls gradually with increasing flow, but for a PD pump, the flow is almost constant whatever the head. It is customary to draw the curve for PD pumps with the axes reversed (see Section 4.3), but to understand the interaction with the system, a common presentation is used here for the two pump types.

2.4 Pump operating point

When a pump is installed in a system, the interaction can be illustrated graphically by superimposing pump and system curves. The operating point will always be where the two curves intersect. (Figure 2.8 and Figure 2.9).

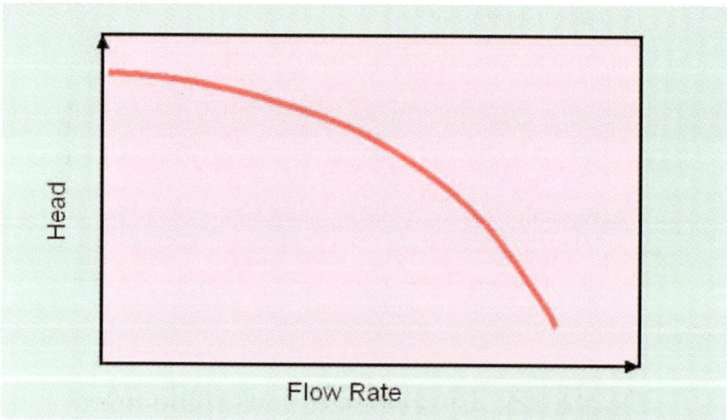

Figure 2.6: Rotodynamic pump

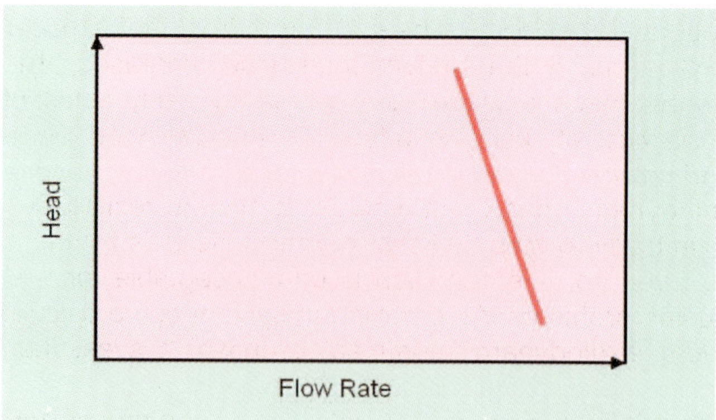

Figure 2.7: Positive displacement pump

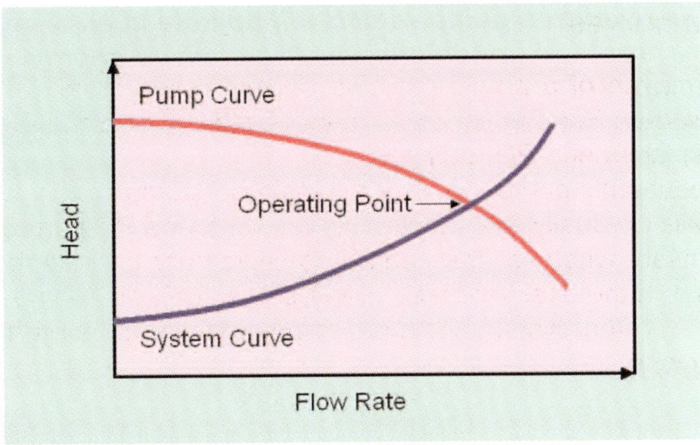

Figure 2.8: Rotodynamic pump and system curves

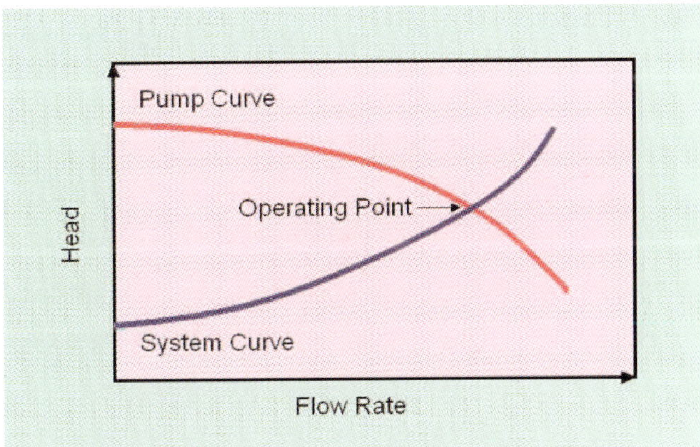

Figure 2.9: PD pump and system curves

If the actual system curve is different in reality from that calculated, the pump will operate at a flow and head different from that expected.

For a PD pump, if the system resistance increases, the pump will increase its discharge pressure and maintain a fairly constant flow rate, dependent on viscosity and pump type. Unsafe pressure levels can occur without relief valves.

For a rotodynamic pump, an increasing system resistance will reduce the flow, eventually to zero, but the maximum head is limited as shown in Figure 2.8. Even so, this condition is only acceptable for a short period without causing problems. An error in the system curve calculation is also likely to lead to a rotodynamic pump selection which is less than optimum.

Adding comfort margins to the calculated system curve to ensure that a sufficiently large pump is selected will generally result in

installing an oversized pump, which will operate at an excessive flow rate or in a throttled condition, which increases energy usage and reduces pump life.

CHAPTER • THREE

System and process requirements

3.1 Supply and demand controlled systems

When designing a pumping system the process requirements of flow and pressure have to be well defined. These requirements will dictate the selection of pumps and the way the pumps are controlled. The need to control pump operation varies depending on whether the pumping system is *supply controlled* or *demand controlled.*

Supply controlled systems pump away incoming flows. They may need to match the rate of variable inflow, and hence they may require variable pump operation. If the inflow varies, one or several constant speed pumps can usually accommodate these variations if some storage volume is provided, however this may not always be the most cost effective approach, an infinitely variable flow may sometimes be more appropriate. Examples of supply-controlled systems are pumping sewage, storm water, discharges of process liquids, and de-watering or flood protection schemes.

Demand controlled systems pump liquids into a process or to recipients at a constant or variable rate, as required. Examples of demand-controlled systems are those pumping drinking water, cooling liquid and process liquid. Again, variations in demand can be handled by variable speed pumps or by using storage in combination with constant speed pumps.

3.2 Introduction to variable speed concept

The use of variable speed drives (VSDs) in pumping systems may be motivated not only by the need to adjust the pump output to the supply or demand, but also by the prospect of economic benefits, primarily related to energy and maintenance savings, and reliability improvements. Other benefits may be a reduction of hydraulic transients, as VSD pumps can be soft started and stopped, a reduction in leakage from a system when it operates at a reduced pressure, and benefits associated with better control of pumping operation. These have to be evaluated against the capital costs associated with the purchase of VSD equipment.

The energy savings are unquestionable in systems dominated by friction losses, such as water circulating systems. In such systems, the pumps operate at a constant efficiency irrespective of the pump speed. However, in systems dominated by static head, the pump efficiency changes with the speed. It is typically reduced at lower speeds and hence energy savings may not always be realised (see Section 4.2.2).

There are also some potential problems with the use of electronic VSDs in pumping systems. One category of problems is related to electromagnetic emissions that must be carefully filtered to avoid interference and ensure compatibility.

Another category of problems is associated with pumping liquids containing solids, such as industrial slurries or wastewater. In order to transport solids in liquids, the mixture has to move at a velocity higher than a certain critical velocity below which the solids are deposited. If solids are allowed to deposit in pipes, the friction losses increase progressively as the available flow area is reduced. Also pumps can clog with solids when they operate at a reduced speed. Pumping under such conditions is not only wasteful in energy, but it may cause complete blockage of the system or pumps and disrupt pumping completely.

When considering various options of controlling pump operation, appropriate considerations have to be given to both the technical functionality of the system as a whole and the costs involved.

VSDs offer many advantages, but their potential drawbacks should not be ignored. See Chapter 11 for a more detailed analysis of the benefits, drawbacks and operational issues of using VSDs.

3.3 Process requirements

In a pumping system, the operating costs can be correctly evaluated and the potential savings identified only by considering that all components of the system are interdependent.

Therefore, when examining or designing a pumping system, the process requirement must first be established under the form of a load profile or duration diagram see Figure 5.1.

Then the output control “strategy” must be selected (use of throttling valves, by-passes or VSDs) in order to match the characteristic curve of the pump to the specified system curve(s), reducing the use of energy and minimizing life cycle cost for the complete system.

In general it can be said that flow control of virtually any type of liquid can be achieved by driving the pump(s) at variable speed, instead of using throttling valves or complex by-pass devices.

However, it must be realised that the use of variable speed drives, although useful to improve the accuracy and the flexibility of process control, does not always achieve the minimum life cycle cost target.

Sections 4.2.2, 4.3.3 and 4.3.4, illustrate this subject and give guidance in selecting the most appropriate control method for pumps and motors when considering the characteristics of the system.

CHAPTER • FOUR

Pumps

4.1 Classification of pumps

The classification used in this guide first considers the principle by which energy is added to the fluid, i.e. by dynamic action or by displacement, and then identifies different pump geometries by which this principle is implemented. This approach therefore relates the classification to the pump itself and not to liquids handled, materials of construction, method of drive, or shaft orientation.

All pumps are divided into the two major categories of rotodynamic and positive displacement. Minor categories are excluded. All the rotodynamic pumps shown are centrifugal. Positive displacement pumps are essentially divided into reciprocating and rotary types, depending on the nature of movement of the pressure-producing members. Each of these major classifications is subdivided into several specific types of commercial importance, see Figure 4.1, p.16.[1]

4.2 Rotodynamic pumps

4.2.1 Pump principles and performance characteristics

A rotodynamic (centrifugal) pump is a dynamic device for increasing the pressure of a liquid. In passing through the pump, the liquid receives energy from the rotating impeller. The liquid is accelerated circumferentially in the impeller, discharging into the casing at high velocity, which is converted into pressure as effectively as possible. The actual shapes of the hydraulic passages of the impeller and the casing are extremely important in order to attain the highest efficiency possible.

[1] The form of this classification acknowledges ANSI/HI Pumps-General Guidelines 9.1-9.5 and Pump Handbook by Karassik, Krutzsch, Fraser and Messina.

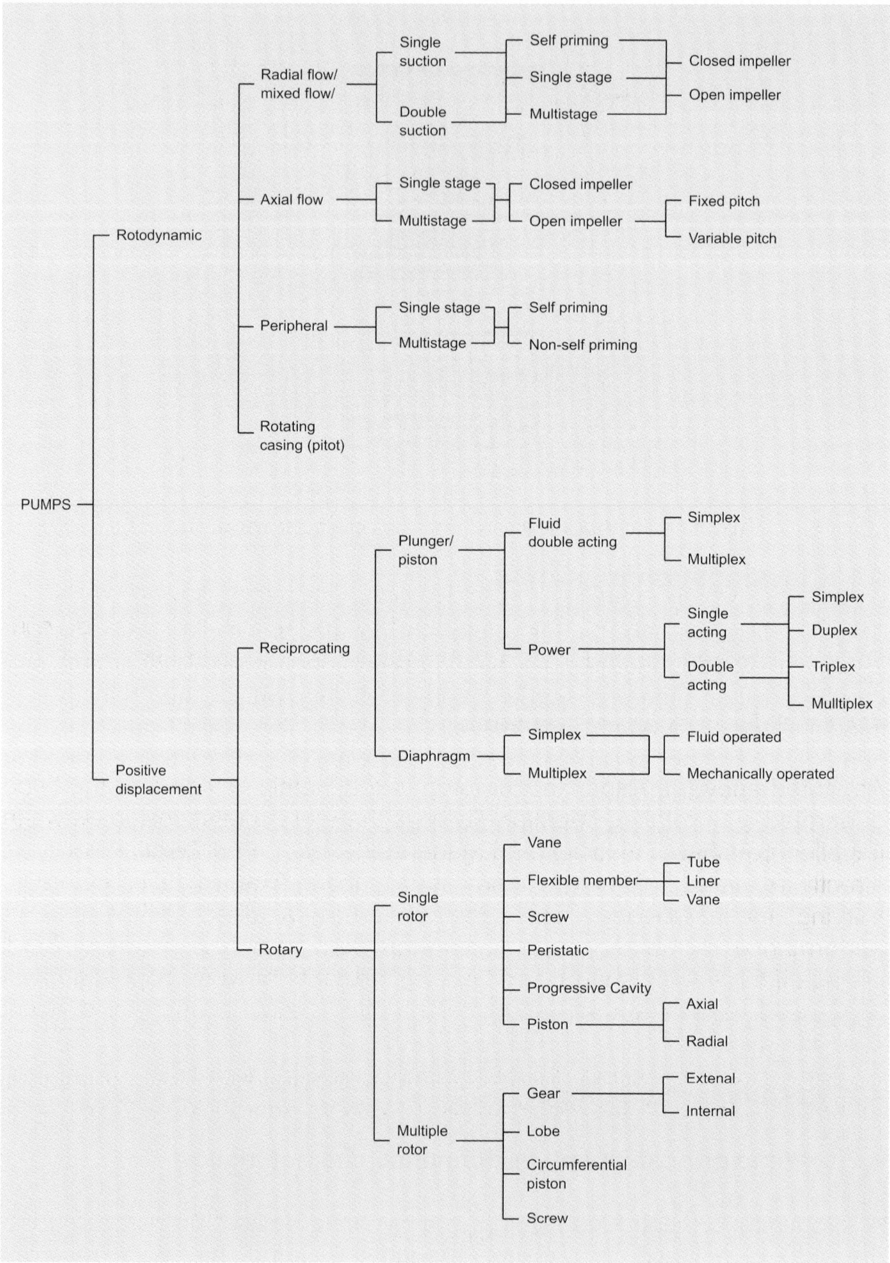

Figure 4.1: Classification of pumps

Since the pump is a dynamic device, it is convenient to consider the head generated rather than the pressure. The pump generates the same head of liquid whatever the density of the liquid being pumped.

The standard convention for rotodynamic pumps is to draw the pump performance curves showing flow rate on the horizontal axis and head

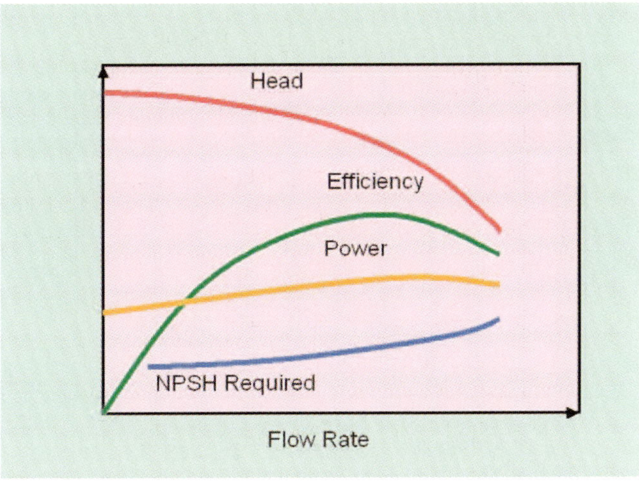

Figure 4.2: Example of pump performance curves

generated on the vertical axis. Hydraulic efficiency, power & NPSH required (see Section 4.2.1.3 for explanation of NPSH), are also all conventionally shown plotted against flow rate, as illustrated in Figure 4.2.

4.2.1.1 Effects of speed variation

As stated above, a rotodynamic pump is a dynamic device with the head generated by a rotating impeller. There is therefore a relationship between impeller peripheral velocity and generated head. Peripheral velocity is directly related to shaft rotational speed, for a fixed impeller diameter. Varying the rotational speed therefore has a direct effect on the performance of the pump. All the parameters shown in Figure 4.2 will change if the speed is varied and it is important to have an appreciation of how these parameters vary in order to safely control a pump at different speeds.

The equations relating the rotodynamic pump performance parameters of flow rate, head and power absorbed, to speed are known as the affinity laws:

$$Q \propto n$$
$$H \propto n^2$$
$$P \propto n^3$$

Where:

Q	=	*Flow rate*
H	=	*Head*
P	=	*Power absorbed*
n	=	*Rotational speed*

Hydraulic efficiency (η) is essentially independent of speed.

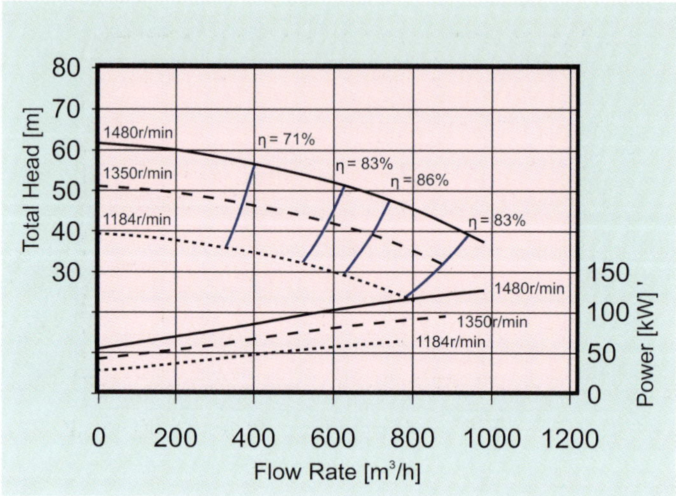

Figure 4.3: Example of speed variation affecting rotodynamic pump performance

The implication of the squared and cubic relationships of head and power absorbed, is that relatively small changes in speed give very significant changes in these parameters as shown in an example of a rotodynamic pump in Figure 4.3.

Points of equal efficiency on the curves for the three different speeds are joined to make the iso-efficiency lines (blue), showing that efficiency remains constant over small changes of speed providing the pump continues to operate at the same position relative to its best efficiency point (BEP).

The affinity laws give a good approximation of how pump performance curves change with speed but in order to determine the actual performance of the pump in a system, the system curve also has to be taken into account, as will be explained later.

The power absorbed by magnetically driven pumps using metallic containment shells includes both hydraulic power and the magnetic coupling power. Where the speed changes are large, or the speed is below 1000 r/min, advice from the pump manufacturer must be obtained. The hydraulic power obeys the affinity laws but the magnetic power absorbed follows a square law with speed. The two types of power must therefore be calculated separately for a change of speed.

This is explained further with a worked example in Appendix A5.

4.2.1.2 Effects of impeller diameter change

Changing the impeller diameter gives a proportional change in peripheral velocity, so it follows that there are equations, similar to the affinity laws, for the variation of performance with impeller diameter D:

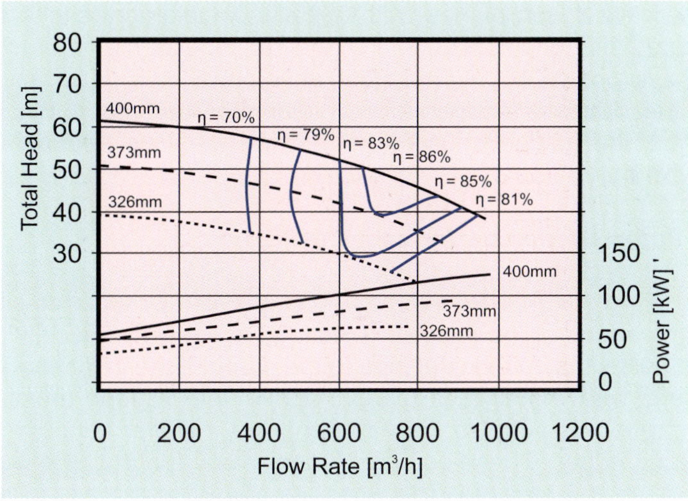

Figure 4.4: Example of impeller diameter reduction affecting rotodynamic pump performance

$$Q \propto D$$
$$H \propto D^2$$
$$P \propto D^3$$

Efficiency varies when the diameter is changed within a particular casing. Note the difference in iso-efficiency lines in Figure 4.3 compared with Figure 4.4. The relationships shown here apply to the case for changing only the diameter of an impeller within a fixed casing geometry, which is a common practice for making small permanent adjustments to the performance of a centrifugal pump. Diameter changes are generally limited to a reduction of about 75% of the maximum, i.e. a head reduction to about 50%. Beyond this, the casing and impeller are mismatched and efficiency and NPSH are badly affected.

In contrast however, speed change can be used over a wider range without seriously reducing pump efficiency as long as the static head is not too high. Motor drive efficiency may however drop.

Certain high flow, low head pumps have performance curve shapes somewhat different from those illustrated, which are typical of radial flow types. Generally mixed flow and axial flow pumps have limited operating regions due to flow instabilities. This requires additional care in matching the pump to the system, when making changes to speed and impeller diameter.[2] (The impeller diameter usually cannot be changed on axial flow pumps.)

[2] Paragraph 1.1.3 & Figure 1.2 of standard ANSI/HI 1.1-2 “Hydraulic Institute Standard—Centrifugal Pumps” is referenced here for further reading on the concept of Specific Speed.

Magnetically driven pumps, with metallic containment shells also need to be treated differently because a change of impeller diameter affects only the hydraulic power, not the magnetic drive power loss. Mechanical power loss in the drive is independent of diameter and so if the speed is unchanged, the magnetic losses will not change. See Appendix A5 for a worked example.

4.2.1.3 Pump suction performance (NPSH)

Liquid entering the impeller eye turns and is split into separate streams by the leading edges of the impeller vanes, an action which locally drops the pressure below that in the inlet pipe to the pump. If the incoming liquid is at a pressure with insufficient margin above the vapour pressure, then vapour cavities or bubbles appear along the impeller vanes just behind their leading edges. These collapse further along the impeller vane where the irpressure is increased. This phenomenon is known as cavitation and has three undesirable effects:

1. The collapsing cavitation bubbles can erode the vane surface, especially when pumping water-based liquids.
2. Noise and vibration are increased, with possible shortened seal and bearing life.
3. The cavity areas will initially partially choke the impeller passages and reduce the pump performance. In extreme cases, total loss of pump-developed head occurs.

The amount by which the pressure in the pump inlet exceeds the liquid vapour pressure is expressed as a head of liquid and referred to as Net Positive Suction Head Available (NPSHA). This is a characteristic of the system design. The value of NPSH needed at the pump suction to prevent the pump from cavitating is known as NPSH Required (NPSHR). This is a characteristic of the pump design.

The three undesirable effects of cavitation described above begin at different values of NPSHA and generally there will be cavitation erosion before there is a noticeable loss of pump head. However for a consistent approach, manufacturers' and industry standards, usually define the onset of cavitation as the value of NPSHR when there is a head drop of 3%, when compared with the head with cavitation free performance. At this defined onset point, incipient cavitation is present and prolonged operation in this condition will usually lead to damage. It is therefore desirable to apply a margin by which NPSHA should exceed NPSHR. The Hydraulic Institute and EUROPUMP have developed guidelines for NPSHR margins. The web sites for the Hydraulic Institute and EUROPUMP are given at the bottom of this page.[3]

[3] Hydraulic Institute online = www.pumps.org EUROPUMP online = www.europump.org

As would be expected, the NPSHR increases as the flow through the pump increases, see Figure 4.2. In addition, as flow increases in the suction pipework, friction losses also increase, giving a lower NPSHA at the pump inlet, both of which give a greater chance that cavitation will occur.

NPSHR also varies approximately with the square of speed in the same way as pump head. Conversion of NPSHR from one speed to another can be made using the following equations.

$$Q \propto n$$
$$NPSHR \propto n^2$$

It should be noted however that at very low speed there is a minimum NPSHR level, NPSHR does not tend to zero at zero speed. It is therefore essential to carefully consider NPSH in variable speed pumping.

The effect of impeller diameter reduction on NPSHR, however, is not the same as on pump head. Pump NPSHR is largely a characteristic of the impeller inlet design, which is not affected by changing the diameter. However because reducing the impeller diameter affects the length of the vanes and the overlap of adjacent vanes, NPSHR increases slightly as the diameter is reduced. The effect is greatest on high specific speed (mixed flow) machines. The pump manufacturer should be consulted on the actual NPSHR variation with impeller diameter.

4.2.2 Methods of varying the performance of rotodynamic pumps

4.2.2.1 The need for performance variation

It has been explained in Chapter 3 that many pumping systems require a variation of flow or pressure. This section describes methods available to do this. Either the system curve or the pump curve must be changed to get a different operating point. Where a single pump has been installed for a range of duties, it will have been sized to meet the greatest output demand, it will therefore usually be oversized, and will be operating inefficiently for other duties. There is therefore an opportunity to achieve an energy cost saving by using control methods which reduce the power to drive the pump during the periods of reduced demand. Not all control methods achieve this goal as explained in this section. In Chapter 5 the cost benefits of different control methods are described in more detail and in Chapter 8 control parameters are elaborated.

Varying pump performance by changing speed is explained first and in many cases is a cost effective approach with good pay back even if the capital expenditure is relatively high. Other methods of control are then explained so that the most appropriate approach, to minimize life cycle cost, can be chosen.

To make an effective evaluation of which control method to use, all of the operating duty points, and their associated run times and energy consumptions, have to be identified so that the total costs can be calculated and alternative methods compared.

Changing pump impeller diameter also effectively changes the duty point in a given system, see Section 4.2.1.2, and at low cost, but this can be used only for permanent adjustment to the pump curve and is not discussed further as a control method.

4.2.2.2 Pump control by varying speed

To understand how speed variation changes the duty point, the pump and system curves are over-laid. It is important to note that the operating point always falls at the intersection of the pump curve and the system curve. Two systems are considered, one with only friction loss and another where static head is high in relation to friction head. It will be seen that the benefits are different.

4.2.2.2.1 Varying speed in a system with only friction head In Figure 4.5, reducing speed in the friction loss system moves the intersection point on the system curve along a line of constant efficiency. The operating point of the pump, relative to its best efficiency point, remains constant and the pump continues to operate in its ideal region. The affinity laws are obeyed which means that there is a substantial reduction in power absorbed accompanying the reduction in flow and head, making variable speed the ideal control method for systems where friction loss predominates.

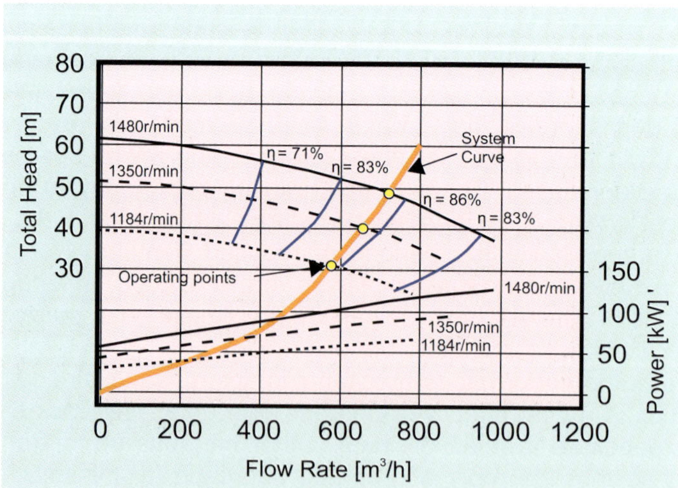

Figure 4.5: Example of the effect of pump speed change in a system with only friction loss

4.2.2.2.2 Varying speed in a system with high static head However, in systems with static head, the system curve does not start from the origin but at some value on the y-axis corresponding to the static head, hence the system curve does not follow the curves of constant efficiency, instead, it intersects them. The reduction in flow is no longer proportional to speed; a small turn down in speed could give a big reduction in flow rate and pump efficiency (see Figure 4.6). In turn, this means that the pump efficiency changes when the speed of the pump changes. This is quite contrary to a pump system without static head, in which the pump efficiency remains fairly constant with changing speed. A common mistake is also to use the affinity laws (see Section 4.2.1.1 for information on affinity laws) to calculate energy savings in systems with static head. Although this may be done as an approximation, sometimes it can also lead to major errors.

As can be seen in Figure 4.6, the operating point on the reduced speed curves also moves relatively higher on the pump curve as the speed is reduced. There are problems connected with this situation. A relatively minor decrease in speed can move the pump operating point into an area where the pump should not be operated continuously. Operating the pump close to shut-off head for extended periods of time could severely damage the pump – even though the pump runs at a reduced speed. Decreasing the speed even further causes the flow to go to zero (the 1184 r/min head/flow curve in Figure 4.6). The pump is still consuming energy, which goes solely into heating the liquid through friction, with the result that damaging temperatures can quickly be reached.

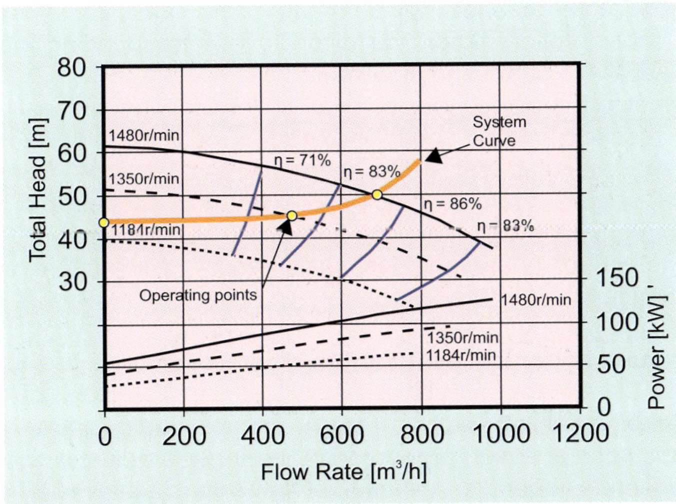

Figure 4.6: Example of the effect of pump speed change in a system with high static head

In the satisfactory speed range with this type of system, the drop in pump efficiency reduces the economic benefits of variable speed control. Systems should be examined on a case-by-case basis. This is discussed further in Chapter 5.

Many times it can be advantageous to select the pump such that the system curve intersects the full speed pump curve slightly to the right of pump best efficiency point. In such a case the pump efficiency will first increase as the speed is reduced and then decrease. This can extend the useful range of variable speed operation in a system with static head. On high-energy pumps particularly, the pump manufacturer should be consulted on the safe operating range of the pump.

It is relevant to note that flow control by speed regulation is always more efficient than by a control valve.

4.2.2.2.3 Other implications of varying speed In addition to energy savings there could be other benefits of lower speed. The hydraulic forces on the impeller, created by the pressure profile inside the pump casing, reduce approximately with the square of speed. These forces are carried by the pump bearings and so reducing speed increases bearing life. It can be shown that for a rotodynamic pump, bearing life is proportional to the seventh power of speed. In addition, vibration and noise are reduced and seal life is increased providing the duty point remains within the allowable operating range.

The corollary to this is that small increases in the speed of a pump significantly increase power absorbed, shaft stress and bearing loads. It should be remembered that the pump and motor must be sized for the maximum speed at which the pump set will operate. At higher speed the noise and vibration from both pump and motor will increase, although for small increases the change will be small. If the liquid contains abrasive particles, consideration should be given to a corresponding increase in surface wear in the pump and pipework.

The effect on the mechanical seal of the change in seal chamber pressure, resulting from a speed increase, should be reviewed with the pump or seal manufacturer, if the speed increase is large. Conventional mechanical seals operate satisfactorily at very low speeds and generally there is no requirement for a minimum speed to be specified. However due to their method of operation, gas seals require a minimum peripheral speed of 5 m/s.

4.2.2.3 Pumps in parallel switched to meet demand

Another energy efficient method of flow control, particularly for systems where static head is a high proportion of the total, is to install two or more pumps to operate in parallel. Variation of flow rate is achieved by switching on and off additional pumps to meet demand. The combined pump curve is obtained by adding the flow rates at a specific head. The head/flow rate curves for two and three pumps in parallel are shown in Figure 4.7.

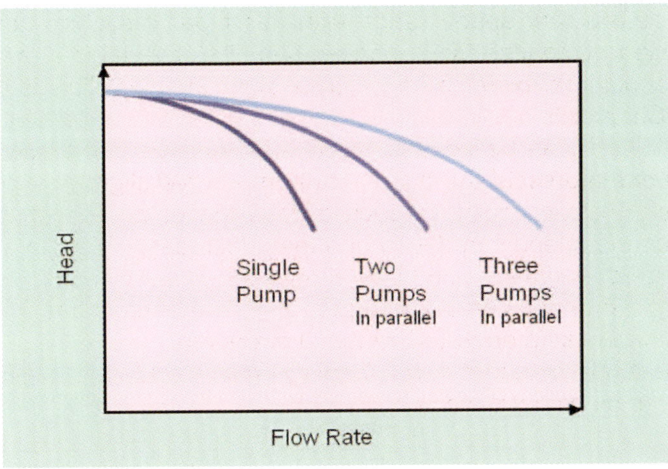

Figure 4.7: Typical curves for pumps in parallel

The system curve is usually not affected by the number of pumps that are running. For a system with a combination of static and friction head loss, it can be seen, in Figure 4.8, that the operating point of the pumps on their performance curves moves to a higher head and hence lower flow rate per pump, as more pumps are started. ***All pumps operate at the pressure where the combined pump curve and the system curve intersect.*** It is also apparent that the flow rate with two pumps running is not double that of a single pump. If the system head is only static, then flow rate is proportional to the number of pumps operating.

It is possible to run pumps of different sizes in parallel providing their closed valve heads are similar. By arranging different combinations of

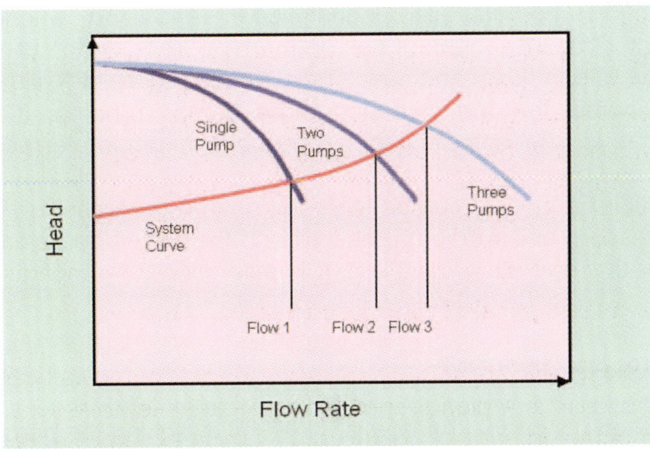

Figure 4.8: Typical curves for pumps in parallel, with a system curve

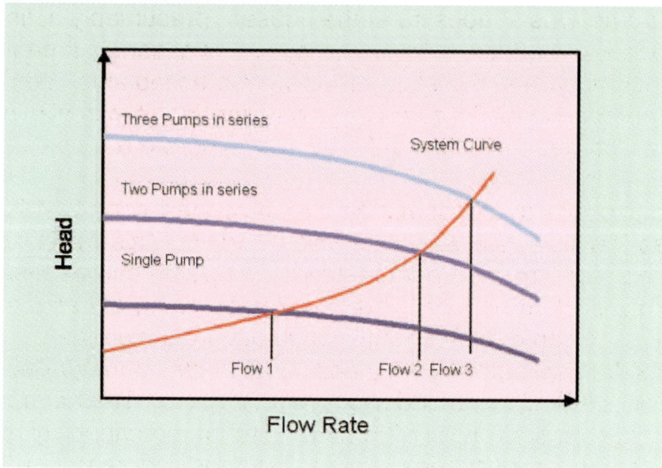

Figure 4.9: Typical curves for pumps in series, with a system curve

pumps running together, a larger number of different flow rates can be provided into the system.

Care must be taken when running pumps in parallel to ensure that the operating point of the pump, when only one pump is running, is acceptable. It can be seen from Figure 4.8 that the pump operates well out along the curve, where NPSHR and vibration levels are higher. The pump manufacturer should confirm that this is a satisfactory operating condition for the period of time that it occurs.

4.2.2.4 Pumps in series switched to meet demand

Rotodynamic pumps may also be installed in series and switched to meet the demand. This arrangement can be more appropriate to systems with a high friction head loss and low static head. Figure 4.9 illustrates the operating points for one, two and three pumps running. It should be verified that all the operating points remain in the preferred operating region around pump best efficiency.

The second and third pumps in series have a progressively increasing suction pressure and working pressure and they must be selected appropriately for this operating condition. It is also difficult to configure the pipe system to vary the 'lead' pump.

4.2.2.5 Stop/start control

In this method of control, the flow is varied by switching the pump on or off. It is necessary to have a storage capacity in the system e.g. a wetwell, an elevated tank or an accumulator type pressure vessel. The storage can provide a steady flow to the system with an intermittently operating pump.

When the pump runs, it does so at the chosen (presumably optimum) duty point and when it is off, there is no energy consumption. If intermittent flow, stop/start operation and the storage facility are acceptable, then this is an effective approach to minimize energy consumption. It may also be used to benefit from "off peak" energy tariffs by arranging the run times during the low tariff periods.

To minimize energy consumption with stop/start control, it is better to pump at as low flow rate as the process permits. This minimizes friction losses in the pipe and an appropriately small pump can be installed. For example, in a system dominated by friction, pumping at half the flow rate for twice as long can reduce energy consumption to a quarter.

The stop/start operation causes additional loads on the power transmission components and increased heating in the motor. The frequency of the stop/start cycle must be within the motor and pump capability and will be limited by the inertia of the load, the power and the speed of the pump, and the quality of the power supply.

4.2.2.6 Flow control valve

With this control method, the pump runs continuously and a valve in the pump discharge line is opened or closed to adjust the flow to the required value.

To understand how the flow rate is controlled see Figure 4.10. With the valve fully open, the pump operates at Flow 1. When the valve is in the half open position it introduces an additional friction loss in the system. The new system curve cuts the pump curve at Flow 2, which is the new operating point. The head difference between the two curves, at Flow 2, is the pressure drop across the valve.

It is usual practice with valve control to have the valve 10% shut even at maximum flow. Energy is therefore wasted overcoming the resistance through the valve at all flow conditions.

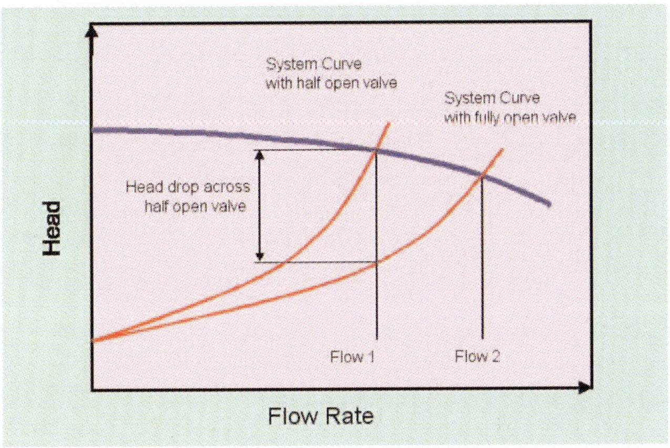

Figure 4.10: Control of pump flow by changing system resistance using a valve

There is some reduction in pump power absorbed at the lower flow rate (see Figure 4.2), but the flow multiplied by the head drop across the valve, is wasted energy. It should also be noted that, whilst the pump will accommodate changes in its operating point as far as it is able within its performance range, it can be forced to operate high on the curve where its efficiency is low (see again Figure 4.2), and where its reliability is impaired.

Maintenance cost of control valves can be high, particularly on corrosive and solids-containing liquids, and so the life cycle cost of this widely used method of control can be surprisingly high.

4.2.2.7 By-pass control

In this approach, the pump runs continuously at the maximum process demand duty, with a permanent by-pass line attached to the outlet. When a lower flow is required the surplus liquid is bypassed and returned to the supply source. An alternative configuration may have a tank supplying a varying process demand, which is kept full by a fixed duty pump running at the peak flow rate. Most of the time the tank overflows and recycles back to the pump suction. This is even less energy efficient than a control valve because there is no reduction in power consumption with reduced process demand.

The small by-pass line sometimes installed to prevent a pump running at zero flow is not a means of flow control, but required for the safe operation of the pump.

4.3 Positive displacement pumps

Positive displacement pumps can be classified into two main groups: rotary and reciprocating.

Rotary pumps typically work at pressures up to 25 bar. These pumps transfer liquid from suction to discharge through the action of rotating rotors, screws, lobes, gears, rollers, etc. which operate within a rigid casing. Rotary pumps do not normally require non-return or check valves, in order to function.

Reciprocating pumps typically work at pressures up to 500 bar. These pumps discharge liquid by changing the internal volume. Reciprocating pumps normally require non-return or check valves fitted to suction and/or discharge, these are usually integral with the pump construction.

Flow rates of PD pumps are in general directly proportional to speed, making them ideally suited to flow adjustment by means of variable speed.

4.3.1 Rotary positive displacement pumps

4.3.1.1 Flow rate, pressure and speed

The relationship between flow rate, pressure and speed varies according to pump type and viscosity of pumped fluid. For pumps having a fall off in flow rate with increasing pressure, the flow rate discrepancy is referred to as 'slip flow', see Figure 4.11. Rotary PD pumps with minimal slip flow can create very high, possibly dangerous pressures if the system is incorrectly selected or a valve on the discharge is closed or even nearly closed. This occurs because the characteristic of this type of pump is constant flow, which causes the pressure to rise to force the flow through the tightest restriction. If the discharge flow can be reduced to zero, safely, by increasing the pressure from P_1 to P_2 then this is referred to as 'dead heading'.

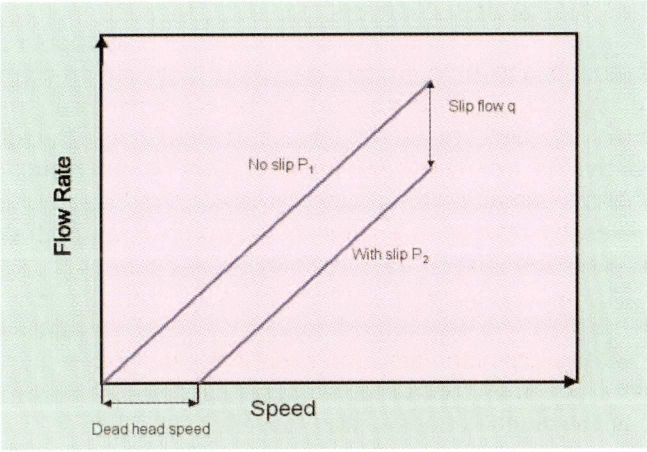

Figure 4.11: Slip flow

Considering the effect of speed change on volumetric efficiency, for the particular example shown in Figure 4.12, it can be demonstrated that by reducing speed from n_1 to n_2 whilst maintaining a constant pressure P_2, volumetric efficiency is reduced from 70% to 50%. Volumetric efficiency is

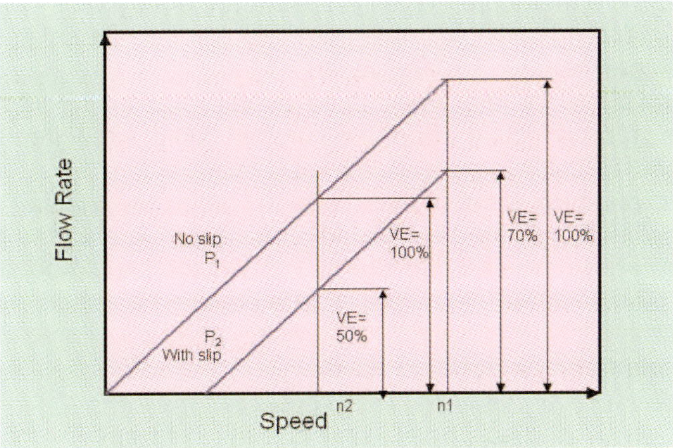

Figure 4.12: Volumetric efficiency

the ratio of volume delivered to displaced volume. When there is no slip, the volumetric efficiency is 100% and the flow rate versus speed relationship follows the no-slip line.

Unlike rotodynamic pumps, changing system resistance i.e. by closing a valve cannot control the flow rate of most PD pumps. This simple method is both ineffective and potentially unsafe as these pumps have the potential to generate very high pressure, under such circumstances.

4.3.1.2 Suction performance

In a similar way to that described in 4.2.1.3, a PD pump needs the incoming liquid to have a pressure margin above liquid vapour pressure to prevent cavitation within the suction passages or pumping chamber. For a PD rotary pump, the required pressure at the inlet is referred to as NPSHR (net positive suction head required). NPSHR characteristics are determined by the geometry and construction of the design, with little ability to change without major modifications.

The cavities on a positive displacement pump open and close as the pump operates and the liquid must flow quickly enough at the pump suction to fill the cavities completely with liquid. If the cavities do not fill with liquid, there will be a void, and apparent cavitation, resulting in a loss of both flow and pressure performance. If the speed is increased, this filling process becomes more problematical, limiting the maximum speed. With viscous liquids, the effect is more severe.

4.3.1.3 Speed and torque

Rotary PD pumps have a constant torque characteristic in contrast to that of rotodynamic pumps whose torque required is proportional to speed squared.

Although the torque required to drive a PD pump, in general, is directly proportional to differential pressure and independent of speed, the torque required to drive the pump across a speed range may vary, depending on the pump operating principle, materials of construction, running clearances and the viscous behaviour of the liquid.

Some rotary PD pump types have a high starting torque. This may result from start-up friction where the pumped liquid is required to lubricate moving mechanical components, or may occur if the pump employs an elastomeric pumping element that displays hysteresis. Starting torque can be significantly more than the normal running torque and can be a major factor in sizing a drive (see Section 8.3.1). Advice should be sought from the pump manufacturer for any application beyond published data.

4.3.1.4 Speed and power absorbed

Typical power absorbed curves vary linearly with speed and pressure; examples are shown in Figure 4.13. Viscosity is assumed constant in the

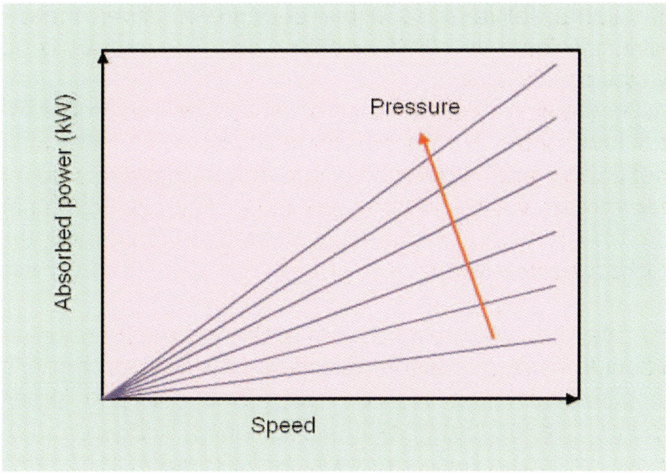

Figure 4.13: Typical power curves

illustration; an increase in viscosity will increase power absorbed. Again advice should be sought from the pump manufacturer.

4.3.2 Reciprocating positive displacement pumps

4.3.2.1 Principles

Reciprocating pumps can generally be classified as having a piston, plunger or diaphragm, displacing a discrete volume of liquid between an inlet valve and a discharge valve.

The rotary motion of the driver, such as an electric motor, is converted to the reciprocating motion by either a crankshaft, camshaft or swash-plate.

The pump can have a number of pumping elements, e.g. plungers, generally between 1 and 7.

The simplest form is a single plunger usually referred to as a single cylinder or simplex pump.

For reciprocating positive displacement pumps, inlet requirements are better expressed by Net Positive Inlet Pressure (NPIP) rather than NPSH.

4.3.2.1.1 Simplex pump There are many more components in such a pump compared to those found in a rotodynamic pump. e.g. crankcase and cover, crankshaft with bearings and seals, connecting-rod, plunger-rod, plunger, seals, and a pump cylinder head with inlet and discharge valve assemblies.

Figure 4.14 shows the plunger displacement with crankshaft rotation. From 0 degrees, bottom dead centre, to 180 degrees, top dead centre, flow is taking place. The area under the curve gives the total volume displaced. The Y-axis also represents how flow rate varies with crankshaft rotation. Between 180 degrees and 360 degrees, filling of the cylinder is taking place but there is no discharge flow. The output is a series of sinusoidal humps.

In practice there will be a short delay to the start of the 'hump' as clearance is taken up between the connecting rod and the crankshaft; also the discharge valve will be delayed in opening as the pressure increases in the cylinder and the valve inertia is overcome. There will be a similar effect at the end of the 'hump' as mechanical clearance is taken up, the pressure difference across the inlet valve is generated and the inlet valve inertia is overcome.

The flow rate Q is a function of time Q(t) and rises from 0% to 315% of the mean flow in a quarter of the revolution. In the next quarter revolution, pumping continues, but the flow rate falls from 315% of the mean to 0%.

In the second half of the revolution nothing is being pumped as during this time the cylinder re-fills. Pressure will generally also vary with time because the flow rate is so uneven.

Even if the pump is delivering to a constant pressure, the torque requirement from the driver has the same large fluctuation of up to 315% of the mean and follows the same profile as the flow rate in Figure 4.14.

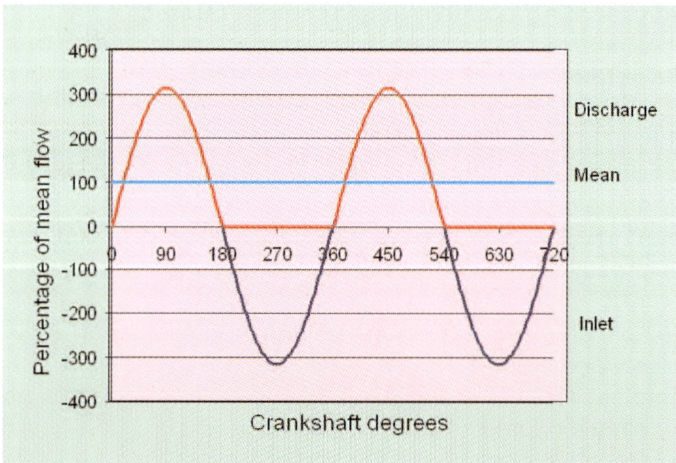

Figure 4.14: Output characteristic of a single cylinder pump

4.3.2.1.2 Duplex pump Adding another set of parts: con-rod, plunger rod, plunger and seals, inlet and discharge valve assemblies, creates a two cylinder or duplex pump. The second plunger is set 180 degrees out of phase with the first. This gives a degree of improvement over a simplex pump, but there is still a wide variation in the flow rate, and hence the torque, see Figure 4.15. The fluctuation still goes from 0% but only reaches 157% of the mean. The frequency is 2 x (r/min)/60 Hz.

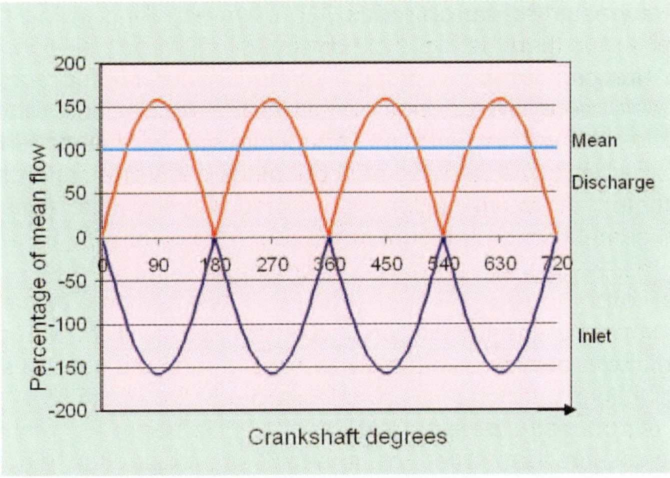

Figure 4.15: Output characteristic of a two-cylinder pump

4.3.2.1.3 Triplex pump Adding a third set of parts makes a dramatic difference, because for two thirds of the stroke there are two plungers discharging at the same time and for the remaining one third one plunger is near to maximum output. The three outputs add together to give a smoother waveform. This is shown in Figure 4.16. Although this is theoretical, a well-designed pumping system will exhibit this waveform albeit being somewhat less smooth.

The flow and hence torque variation is from 90% to 104% of the mean, a total of 14%.

Adding further sets of parts, effects a more limited improvement, which only becomes cost effective for special duties.

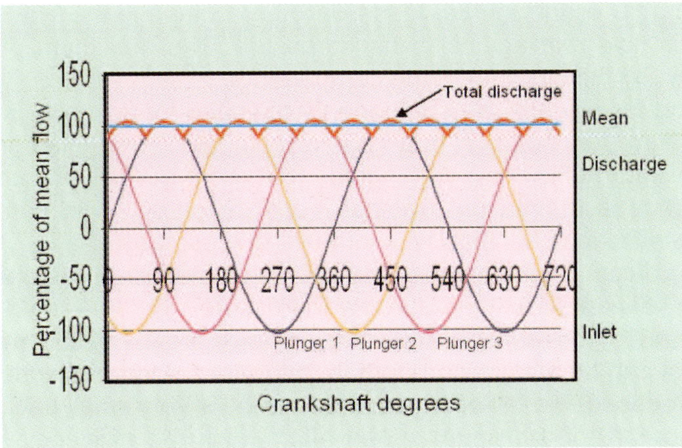

Figure 4.16: Flow characteristic of a three-cylinder pump

4.3.2.2 Reciprocating pump types

4.3.2.2.1 Plunger pump A plunger pump is differentiated from a piston pump in that the plunger moves past a stationary packing, whereas the piston carries the packing with it. With a modern, well designed, plunger pump there is negligible leakage past the seals and valves affording a very high volumetric efficiency, and a total efficiency (volumetric x mechanical) that can exceed 90%, on a triplex pump.

4.3.2.2.2 Piston pump Piston pumps have similar efficiencies to plunger pumps, but are limited to a maximum pressure of 100 bar. This is because of the problems associated with supporting the moving seal.

4.3.2.2.3 Diaphragm pump Where the diaphragm is displaced by hydraulic pressure generated by a seal-less plunger the diaphragm is evenly loaded and supported. This extends the diaphragm life. However the leakage past the hydraulic plunger decreases the efficiency in a similar way to the slip flow of rotary pumps. This slip flow increases with pressure.

Where the diaphragm is mechanically actuated the diaphragm displacement is less efficient.

4.3.2.3 Effect of speed variation for plunger and piston pumps

These pumps move a discrete volume of liquid on each stroke. They are flow generators. Double the speed and the flow is doubled. The pressure function comes as a secondary result of the restriction to flow that occurs downstream of the pump. The pump has to force this positive flow against the restriction.

A pump has a maximum flow rate (hence r/min). This may be limited by any one of a number of factors: the size of the valves, the limiting speed for bearing life or the inertial loads on other components. The pump will deliver this flow rate against a backpressure. The maximum pressure is determined by the strength of the components.

A design usually has the components matched for flow and pressure at the maximum duty point. With a leak-free plunger pump this corresponds to the maximum efficiency. With liquids that are practically incompressible the efficiency is maintained irrespective of pressure. With compressible liquids the efficiency will drop with increasing pressures.

In older designs where there was slip flow to lubricate the packing, the maximum efficiency could be at a lower pressure.

When the pump is run slowly and at less pressure, the mechanical energy lost can be a higher proportion of the total energy absorbed. When speed is reduced, the fall off in pump efficiency is very small until the pump is down to 1/10 of maximum speed after which the efficiency falls more rapidly.

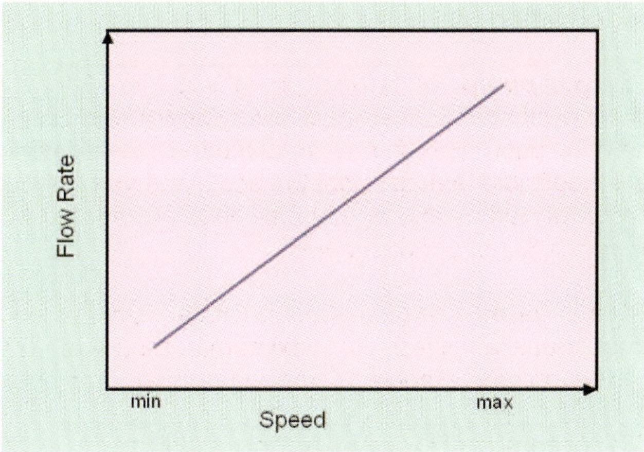

Figure 4.17: Flow rate vs. speed for a plunger pump

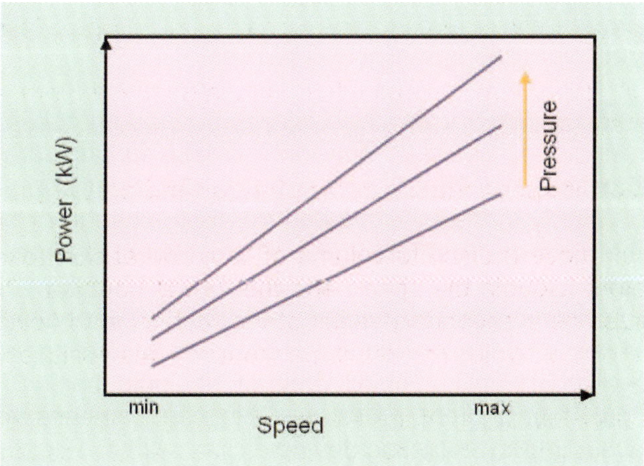

Figure 4.18: Power vs. speed for a plunger pump

Typical performance graphs for plunger pumps, without compressibility, are shown in Figures 4.17 and 4.18.

Figure 4.19 shows typical efficiency for a plunger pump and should be used in conjunction with the Figure A6.4 in Appendix A6 to calculate the overall efficiency of pumps plus motor, at different speeds.

4.3.2.4 Effect of speed variation for diaphragm pumps

Performance curves for diaphragm pumps are similar to those for rotary PD pumps. Due to hydraulic slip flow or mechanical diaphragm distortion, the flow rate is also a function of pressure, losing the proportionality with r/min. The efficiency then drops with increasing pressure.

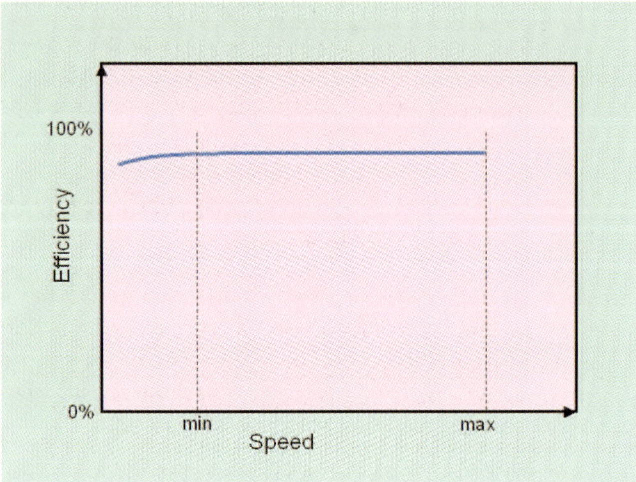

Figure 4.19: Efficiency vs. speed for a plunger pump

4.3.3 Applying variable speed to PD pumps

Throttling the output from a positive displacement pump does not achieve a reduction in pump flow rate, only an increase in pressure. This is a dangerous practice.

To control flow rate, the pump speed needs to change or some of the flow has to be diverted. To effect a one-off permanent change, a modified drive train can be used, (e.g. different pulley sizes). For many applications some small flow rate changes need to be made, whilst holding pressure constant and this is best achieved with a pressure-regulating valve.

Such a valve will spill a small amount of liquid back to source to maintain a constant system pressure. This can compensate for small amounts of wear in any restricting device.

The use of such a valve to spill large volumes of liquid will be very inefficient, the loss of energy manifesting as heat and noise.

For an application where the flow rate needs to vary on a regular basis, a variable speed drive is the preferred option. This is the most efficient method of flow control and it does not waste any of the shaft input energy. The motor will be slightly less efficient at the lower speeds, the 100% T/Tn curve in Figure A6.4.

In a rotodynamic pump, usually only small changes in speed, typically no more than 2:1, are necessary, to make significant energy savings and to move the duty point to much smaller flow rates. This results from the affinity laws, Section 4.2.1.1, where head falls as speed squared and power as speed cubed, and also from the interaction of the pump and system curves, see Section 4.2.2.2.

For a PD pump, the flow rate is proportional to speed, but the pressure can be independent of speed. When the pump is supplying a fixed size of restriction, lowering the flow reduces the pressure in a squared relationship, 50% flow rate resulting in 25% of the pressure. When the pump is supplying at a fixed pressure, e.g. injecting into a high pressure main, then the pressure is constant. Consequently power and energy saving fall linearly when speed reduces. It can become a requirement to operate PD pumps over a wider speed range than rotodynamic pumps, typically up to 10:1.

This large speed range and the characteristics of PD pumps have implications for both the pump and the drive train:

- There may be insufficient lubrication in the crankcase at low speeds.
- Lower or higher operating speeds may require special consideration with respect to the method or type of lubrication and/or cooling.
- The motor may not be adequately cooled at the lowest speed. A separately driven fan may need to be considered.
- The flow rate may be so low that the valve opening is too small to be sustainable under the different forces and the valve could flutter.
- The energy from the drive train inertia becomes too small to smooth the torque ripple and the motor starts to hunt. A motor running at a higher speed with a bigger drive train reduction ratio or a compensating flywheel can be optional solutions.
- At the system design stage, the constant torque characteristic and possible low speed torque effects, must be considered, creating demands on electronic variable speed drives.
- When pumping liquids containing solids, where the solids have a high settling rate, there may be excessive accumulation of solids in the pump and resulting wear. It is paramount when reducing speed with such liquids, that the velocity be maintained high enough, in the pump and in the pumping system, to avoid settling out of the solids.
- A change to liquid properties such as temperature and viscosity could lead to cavitation.
- Viscosity will not affect pump performance until it is sufficiently high to prevent the liquid from entering the cylinder quickly enough. If the cylinder fills, the pump will discharge the volume with no extra power consumption. Power is a function of flow rate and pressure only.
- Liquid compressibility and expansion of the pumping chamber reduces the volumetric efficiency at high pressures.
- The piping system should be designed to avoid resonance at the pump excitation frequencies, over the speed range selected.
- The starting torque requirements will influence the drive selection see Chapter 11.

In most pump systems there is a maximum and minimum speed at which the pump can be run satisfactorily. The pump manufacturer's advice should be sought on this and on the best application of VSDs.

4.3.4 Other methods of flow control for PD pumps

4.3.4.1 Pumps in parallel

PD pumps can generally be run in parallel without problems. This gives the ability to increase flow rate at the pressure rating of a single pump. The principal considerations are, the correct design of inlet and outlet pipework to avoid problems of NPIP/NPSH, overpressure in the discharge pipe and back flow through a non-running pump. Isolation valves for safe maintenance should also be considered.

4.3.4.2 Pumps in series

Rotary pumps in series can give increased pressure capability at the flow rate of a single pump, however it is not recommended that PD pumps be applied in this manner unless sophisticated speed control logic is used. Matching of the two pumps is extremely important and safety, in the event of failure of either unit, normally requires multiple overpressure protection devices and a means of ensuring the second pump is not underfed, creating low NPIP/NPSH and cavitation problems.

Safety of the second pump has to be taken seriously as the system pressure has to be catered for rather than differential operating pressure. The use of one drive for two pumps overcomes many of the starting/speed variation/synchronization issues and compensates for any fluctuations in electrical supply or motor loading.

4.3.4.3 Flow control valve

This is not an acceptable technique for flow control of a PD pump. Throttling the discharge is unsafe as these pumps have the potential to generate very high pressures.

4.3.4.4 By-pass control

Although an effective way of controlling flow rate for small flow variations, it is not preferred for major duties as the waste of energy and possible wear on by-pass control valves give rise to greater running costs. However, it can have the lowest initial cost.

CHAPTER FIVE

Concepts for estimating pumping energy costs

5.1 Flow duration diagrams

When evaluating a pump system it is helpful to be able to identify the system effectiveness easily and to compare different solutions. In order to make an intelligent choice, some basic facts will need to be established. The first is: What are the Process Demands?

The Process Demands will generally indicate a certain direction, i.e. will the process require varying flow rate and, if so, should it be continuously variable or can flow rate be varied in steps? Can on-off batch pumping be used? What is the peak flow rate and how is the flow rate distributed over time?

The answers to these questions will determine if, and how, to regulate the flow. It will also give some guidance regarding the design of the pumping system.

An easy way of showing the flow demand is to use a duration diagram. A duration diagram (see Figure 5.1) in its simplest form shows how many hours (t) during a year that a given flow rate (Q) is needed, the blue line. The red curve in the same diagram is interpreted differently. Each point on the red curve tells how many hours during a year the flow rate exceeds the value on the y-axis.

This diagram is instrumental in understanding the pumping needs. The system must be able to deliver the peak flow but, from an economic point of view, it is also important to know at what flow rates the system is going to operate most of the time.

Given this information, the piping system can be designed. If, for example, the maximum flow rate occurs only for short periods of time, it may not pay to install a large diameter pipe. On the other hand, if the

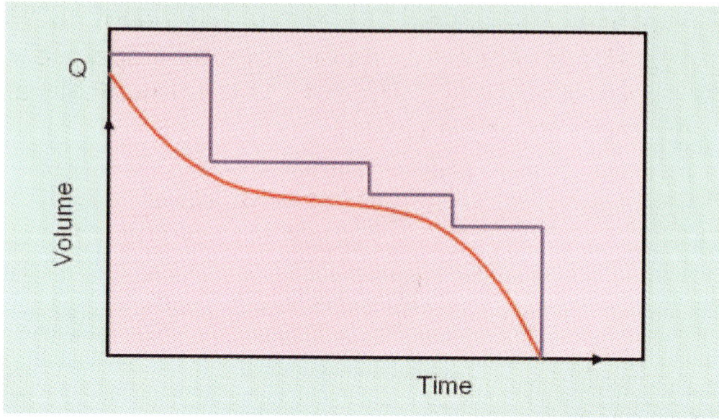

Figure 5.1: Duration diagrams for a pumping system

system is operated at peak flow rates for extended periods of time, this fact should be taken into account when specifying the pipe diameter. When the piping system has been designed, it is time to calculate the system curve.

The system curve (see Figures 2.4 and 2.5) tells how much head, or pressure, is needed from a pump to push a given flow rate through the pipe system.

The duration curve is used to determine where on the system curve the pump will operate and for how long. If a considerable length of time is spent high up on the system curve, a larger pipe diameter, or other modifications, should be considered, so that a pump of lower head and power can be used.

The pump, or pumps, can now be specified, plus suitable means for regulating the flow – if needed. Section 4.2.2.2 explains how a pump performs in conjunction with a variable speed drive.

5.2 Specific energy

A useful measure for calculating the cost of pumping is the specific energy E_S. Specific energy is measured as kWh/m^3, where kW = power in kilowatts, h = time in hours, and m^3 = volume in cubic metres. Specific energy is also a useful measure for comparing different system solutions.

In systems where the flow is constant, this is a simple task by using the equations below. In systems with varying flow rates, it becomes a little more complicated, because each duty must be separately calculated and summated to obtain total costs.

First, E_S needs to be calculated as a function of flow rate. This requires information from pump, motor and drive manufacturers. The pump manufacturer has to provide pump curves for variable speed operation while the motor and drive suppliers have to provide efficiency curves as a function of load and speed.

When E_s has been calculated this information needs to be combined with the duration diagram in order to obtain the total operational cost. The designer can then compare systems with different numbers of pumps as well as different methods of regulation.

$$\textit{Specific Energy } (E_s) = \frac{\textit{Energy Used}}{\textit{Pumped Volume}} \qquad (1)$$

$$E_s = \frac{P_{in} \times Time}{V} = \frac{P_{in}}{Q} \qquad (2)$$

P_{in}= input power to the driver

E_s is a function of flow rate (Q). It is, therefore, necessary to evaluate this dependence. (This is especially important when a variable speed drive is contemplated in order to avoid pitfalls.)

It is useful to separate systems with and without static head since they have different characteristics.

5.2.1 Systems without static head or closed loop systems

The specific energy here is dependent on the frictional head loss which, in turn, is determined by the losses in the pipe system (including throttling valves), and by the combined drive–motor–pump efficiency.

The combined drive–motor–pump efficiency has to be evaluated for each duty point. It is to be noted that the pump efficiency remains approximately the same in a system of this type when the speed is changed, whereas the drive-motor efficiency can drop drastically as the load is reduced. If on the other hand, the system curve is changed by changing the setting of a valve, this will change the duty point of the pump and, hence, its efficiency.

From equation (2) we get

$$E_s = \frac{Q \times H \times \rho \times g}{Q \times \eta_{drive} \times \eta_{motor} \times \eta_{pump}} \qquad (3)$$

Where

H = *Total head*

ρ = *Density*

g = *gravitational constant*

η = *efficiency*

With zero static head, only frictional head H_f remains and thus:

$$E_s = \frac{H_f \times \rho \times g}{\eta_{drive} \times \eta_{motor} \times \eta_{pump}} \qquad (4)$$

5.2.2 Systems with static head

In systems with static head (H_S), the energy usage can be calculated in a slightly different way. The head needed from the pump can be separated into static and dynamic (friction losses). Substituting H_S+H_f for the total head in the expression for specific energy will generate the following expressions:

$$P_{in} = \frac{Q \times (H_S + H_f) \times \rho \times g}{\eta_{drive} \times \eta_{motor} \times \eta_{pump}} \tag{5}$$

$$E_S = \frac{H_S \times H_f}{H_S} \times \frac{\rho \times g \times H_S}{\eta_{drive} \times \eta_{motor} \times \eta_{pump}} \tag{6}$$

if

$$\frac{H_S}{H_S + H_f} = f_{HS} \tag{7}$$

then

$$E_S = \frac{H_S \times \rho \times g}{\eta_{drive} \times \eta_{motor} \times \eta_{pump} \times f_{HS}} \tag{8}$$

The "hydraulic system factor", f_{HS}, indicates the relative amount of static head in the system. It is obvious that E_S now has a minimum value of $H_S \times \rho \times g$ which would occur if all efficiencies were equal to 100% ($\eta = 1$) and there were no friction losses. If there is no variable speed drive in the system, then $\eta_{drive} = 1$. The different factors are all functions of the flow rate and will vary with the duty point. If a variable speed drive is used, they will vary with speed as the duty point moves along the system curve.

Motor efficiency will generally decrease as the speed is lowered and the motor goes below 75% of full load. The drop in combined motor – drive efficiency can be substantial if the motor load drops below 50% of full load. The denominator: $\eta_{drive} \times \eta_{motor} \times \eta_{pump} \times f_{HS}$ can also be seen as the overall (gross) efficiency η_{gr}.

The hydraulic system factor will increase when the friction losses go towards zero, which happens when the duty point approaches the shut off head or when the friction losses are lowered. Hence lowering the friction losses has a substantial effect on the specific energy.

The specific energy will, however, always increase drastically as the duty point moves towards shut off head in systems with static head due to reduced pump, motor and drive efficiencies. In systems with high static head, this can happen even at a relatively moderate decrease in speed.

In such systems the area of usefulness of a variable speed drive can be somewhat improved by making sure that the system curve and the full speed pump curve intersect to the right of the pump's best efficiency point.

To calculate the cost of pumping, the specific energy has to be calculated at all operating points along the system curve. By combining this information with the information in the flow duration diagram the cost of pumping can be determined.

5.3 Flow regulation by varying speed

The resulting curves for specific energy as a function of speed for three different system curves are shown in Figure 5.2. Figure 5.3 shows the corresponding system curves. The intersection of the two reduced speed pump curves with the system curves B and C, indicate shut off head where the specific energy goes towards infinity. It can be seen that the savings potential is very large at low static head, curve A, whereas care has to be taken in high static head situations, represented by curve C.

When the speed is low enough to cause the pump to operate at or close to shut-off head, the specific energy always goes towards infinity. In case A, this will occur since the motor-drive efficiency will approach zero with no load.

The line D indicates the specific energy at which the pump is operating when using on-off control. When operating below this line, energy savings will be realized compared to on-off operation.

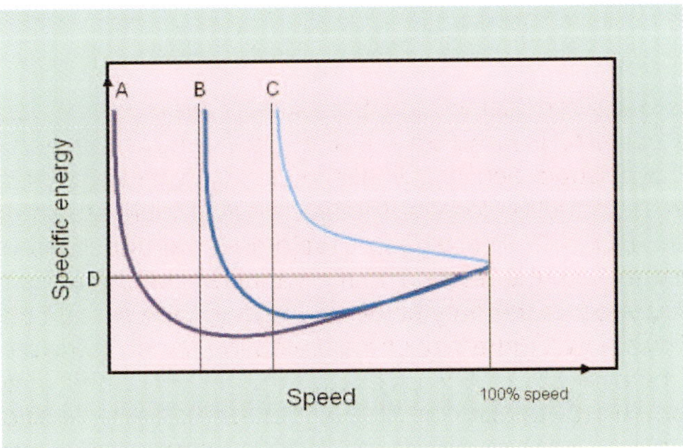

Figure 5.2: Specific energy for three system curves

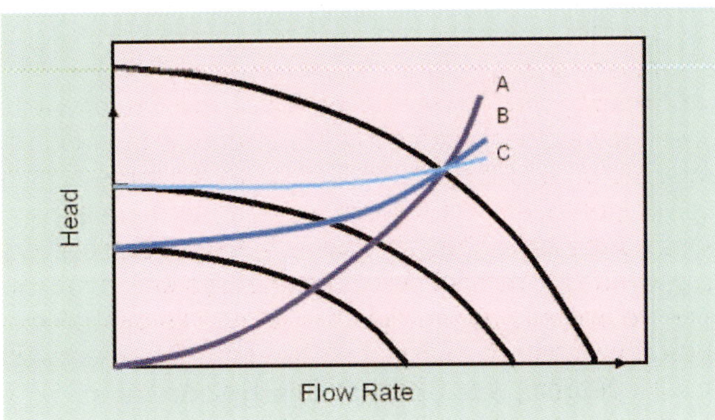

Figure 5.3: System curves for Figure 5.2

5.4 Flow regulated by throttling

When the flow is regulated using a throttling valve, the system curve is changed. The duty point moves to the left on the pump curve, when the flow is throttled, see Figure 5.4. The vertical lines in Figure 5.4 represent the throttling loss in the valve. The specific energy can be calculated for each operating point by dividing the input power to the motor by the flow rate. E_S usually increases rapidly as the flow is reduced, typically like the curve 1 in Figure 5.5.

Compared to regulation by throttling, a variable speed drive *always* saves energy. Line 3 in Figure 5.5 represents the specific energy for an on-off regulated pump.

The specific energy for a speed regulated pump system can be higher than that for an on-off regulated system especially at low flow rates, but will be lower and certainly save energy compared with a throttled system.

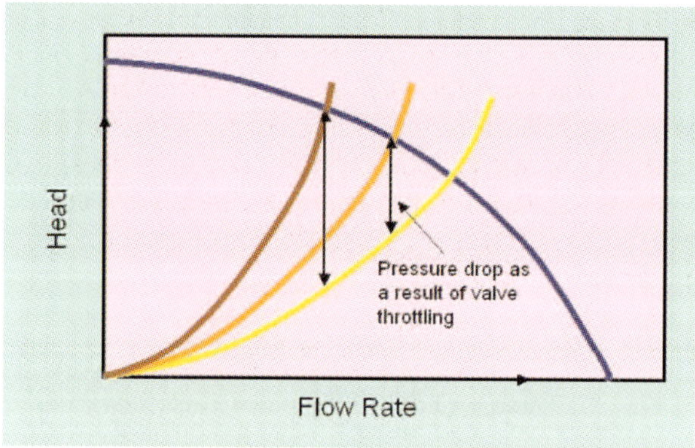

Figure 5.4: Valve throttling losses

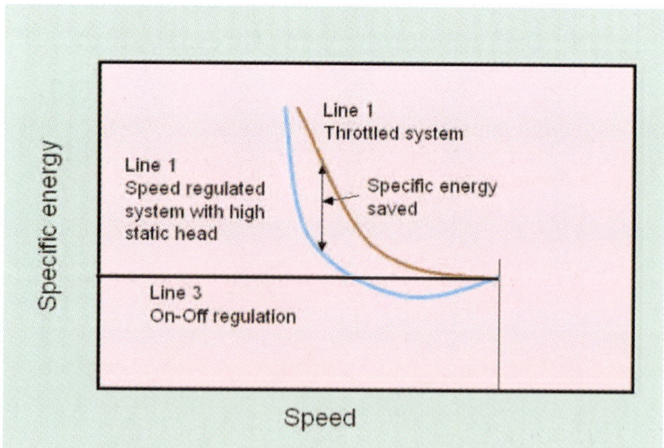

Figure 5.5: Specific energy curves

5.5 Parallel pumps common header

It is important to understand that parallel pumps operating with VSDs and discharging into a common header exhibit the same behavior as previously described when operating a pump against a static head, whether or not static head is present. The first pump discharging to the header pressurizes the header. The second and subsequent pumps that come on line must then pump into a pressurized header. The more pumps that are running the higher the pressure in the header, which limits the chances of saving energy by using a variable speed drive. See Figure 5.6 for a graphical explanation.

With three pumps running, each pump operates at point 3 in Figure 5.6. Thus, a fourth pump must deliver at least this pressure before it produces any flow. This situation is, therefore, identical to pumping against a static head.

This is relevant to water and wastewater schemes where variable speed pumps running in parallel are growing in popularity. Optimization of the variable speed system is essential and specialist advice should be sought. However, as a common rule all the pumps should run to a similar characteristic, which usually means running identical pumps at identical speed. Similarly it is not recommended to run a fixed speed pump in parallel with a variable speed pump of the same size. The potential problem that can arise, and must be avoided, is that one of the pumps operates at no flow.

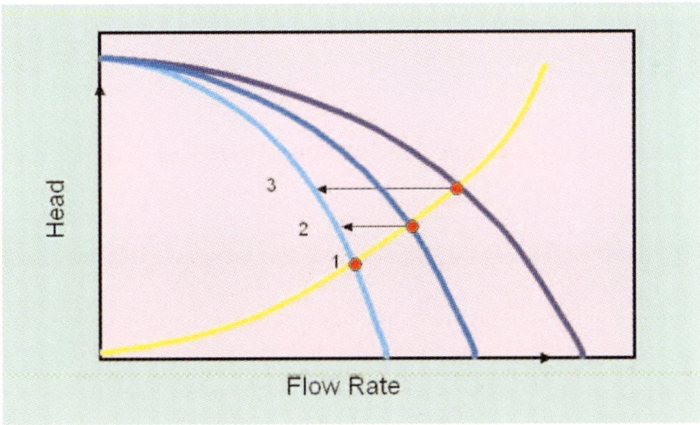

Figure 5.6: Parallel pump operation

5.6 System awareness – notes of caution

To understand a pumping system it must be realized that all of its components are interdependent. Sub-optimization at the component level can easily be deceptive.

The true cost of pumping can be calculated by using the information from Figure 5.1 and Equation (8) where E_S has to be evaluated for all flow rates. In most instances, pumps are sized to deliver the peak flow rate with some margin upwards. With duration curves looking like the continuous curve in Figure 5.1, it is therefore common that the *normal* flow rate is around, or less than, 50 percent of the design flow.

If the system curve exhibits a fair amount of static head and the pump is oversized for most of the pumping needs, then problems can occur. It is not uncommon in such systems to find that the cost of pumping is considerably increased when using a variable speed drive, compared with on-off pumping at full speed. This is generally due to falling pump efficiency at lower flow rates as explained earlier. The fact that the combined efficiency of a variable speed drive-motor package can drop considerably as the load is decreased, does not make the situation better. Thus the allowable speed range becomes restricted both from an operational and economical point of view in such systems. Throttling the flow, however, is still worse!

5.7 Conclusions on a VSD within a system

Variable speed drives will give good control of the flow rate. They generally greatly reduce the operating cost in all systems when compared to throttling-valves. In systems with high relative static head, extra care has to be taken when using variable speed drives, to avoid the pitfalls of low pumping efficiency and operation in harmful flow regimes. In systems with little or no static head VSDs will show reduced operational cost over any other method of flow control.

CHAPTER • SIX

Motors

6.1 Types of electric motors

Whilst there are many types of pump prime movers available, such as diesel engines and steam turbines, the majority of pumps are driven by electric motors.

Electric motors have evolved over the last 150 years, and the most common constructions are detailed in the following pages, and in Figure 6.1.

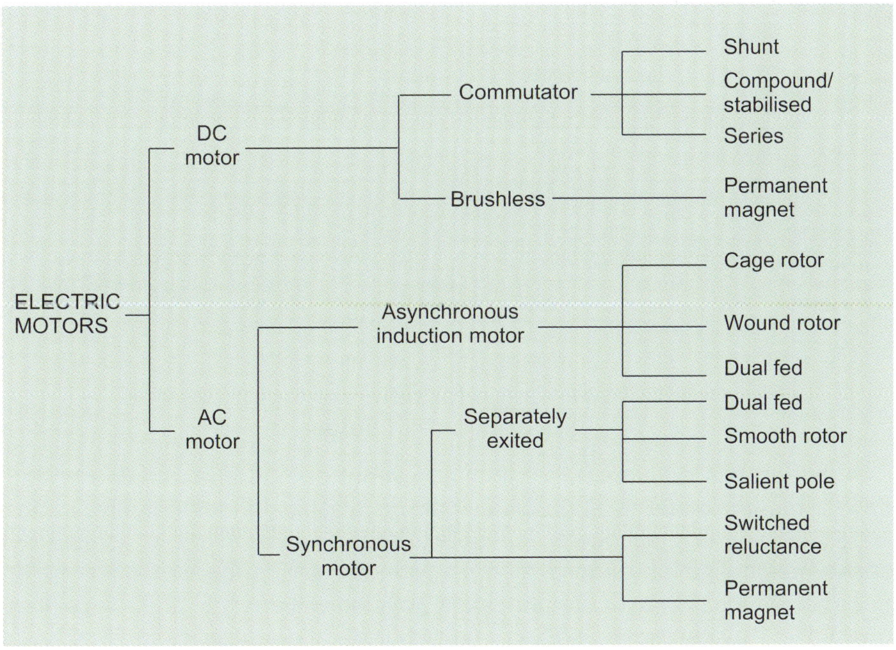

Figure 6.1: Classification of electric motors

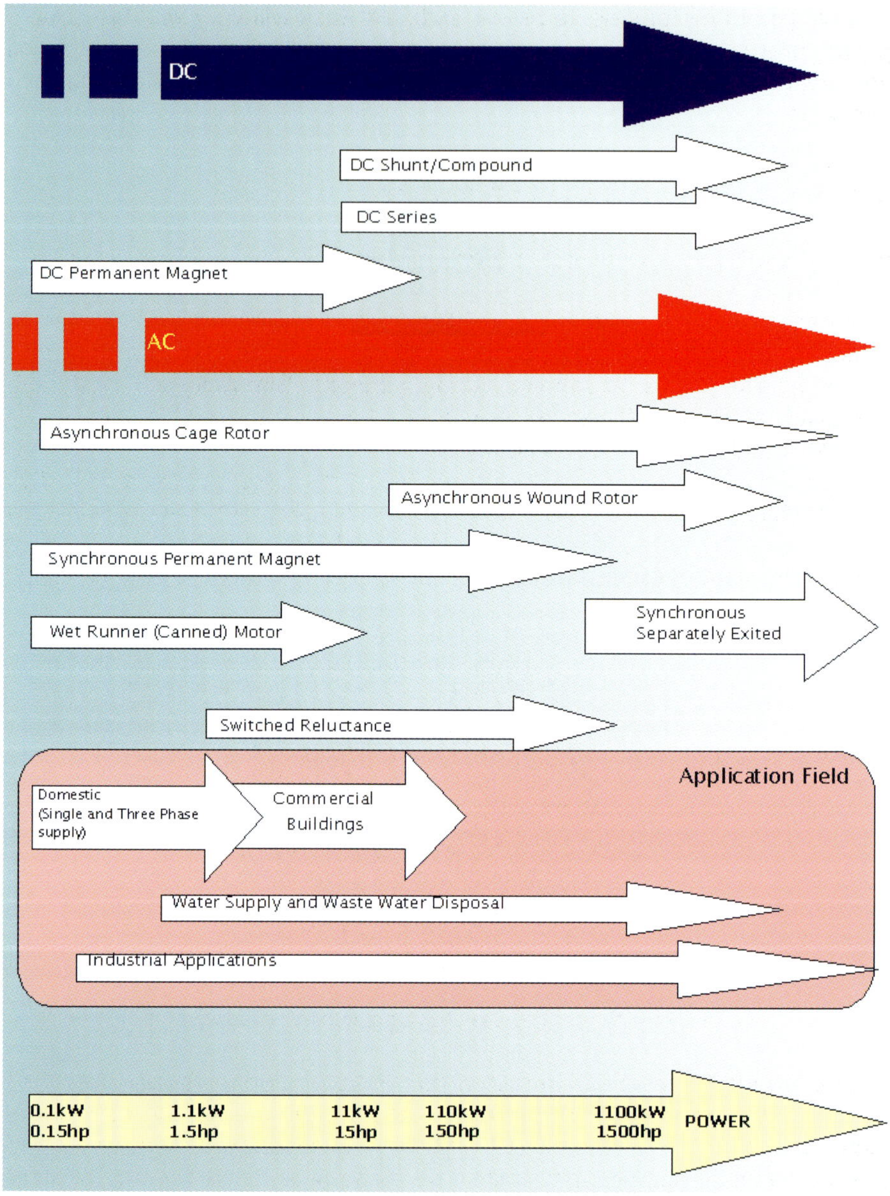

Figure 6.2: Power range for different motor designs, with application fields

Although this document is principally about pumps and variable speed drives it is important to appreciate that, on a typical industrial site, motor driven equipment accounts for approximately two thirds of electricity consumption and any improvements in motor efficiency can offer major energy savings. Many of the principles outlined will apply to all motors on a given site, not just those used as pump drivers whose application is shown in Figure 6.2.

The type of motor selected for a particular task will vary with the speed, power and supply available, whereas the construction of the motor will vary with the environment of the installation and the cooling medium available.

6.2 Asynchronous induction motors

6.2.1 Main types and operating principles

The cage induction motor is by far the most widely used type of electric motor in industry. It is both rugged and reliable; it is also the preferred choice for most variable speed drive applications. Simplicity, low cost, high reliability, fairly high efficiency, coupled with its ease of manufacture, make it widely used in most parts of the world.

Figure 6.3 shows the typical arrangement of a *squirrel-cage induction motor*, which is built with three sets of stator windings arranged around the stator core. The rotor has conductors, which are a cage of copper or aluminium bars and short-circuiting end rings. There are no electrical connections to the rotor.

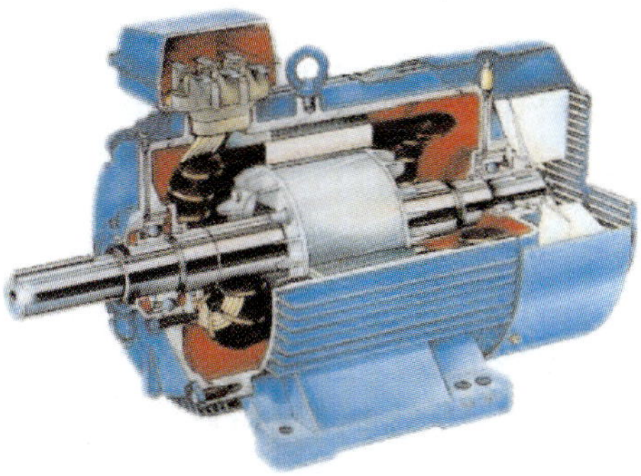

Figure 6.3: Cage induction motor section

The *wound-rotor induction motor,* as the name suggests, has insulated copper windings in the rotor similar to those in the stator. The rotor windings are connected to starting equipment using slip rings and brushes, and therefore this design is substantially more costly, and gives more maintenance problems than a squirrel-cage rotor. This type of induction motor was formerly used in many industrial applications where the starting current and torque needed to be controlled. There are also a number of methods of achieving speed control with a wound rotor machine. However, in new applications, squirrel cage motors are by far the most widely used solution.

For all induction motors the basic principle of operation is that a three-phase voltage supply applied to the stator windings results in the creation of a magnetic field that moves around the stator – a rotating magnetic field. The moving magnetic field induces currents in the rotor conductors, in turn creating the rotor magnetic field, see Figure 6.4.

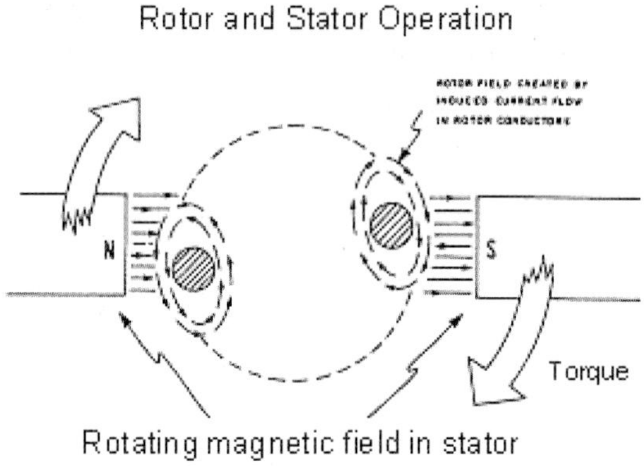

Figure 6.4: Development of torque

Magnetic forces in the rotor follow the stator magnetic field, producing a motor torque. The speed of an induction motor is determined by the frequency of the power supply, by the number of poles in the stator winding and to a smaller extent by the motor load.

The standard construction of a cage rotor motor is designed for use with a symmetrical three-phase electricity supply. The three phases are displaced by 120° electrically, which provides an inherent rotation of the internal magnetic field. It is also possible to make alternative constructions, the most common being a *single phase motor*. In this case a phase displacement is usually achieved by means of a capacitor which provides a 90° basic phase shift, and the winding displacement is based on this value.

An asynchronous motor powered from the mains, has essentially a constant output speed, which decreases only a few percent, typically 1 to 3% (known as ***slip),*** when the motor goes from no-load to full load operation. Therefore, to control the motor speed without the use of external mechanical devices, it is necessary to control the supply voltage and frequency.

The no-load, or synchronous speed for a motor is defined by the formula:

$$n = 120 \times f / p$$

Where

n = speed in r/min
f = supply frequency (Hz)
p = number of poles

Some types of motors may also be manufactured with multiple windings to achieve two or more discrete speeds (see Table 6.1).

6.2.2 Motor efficiency

The majority of induction motors are designed and manufactured to meet design standards, which specify output power, synchronous speed and critical dimensions.

Table 6.1: No load (synchronous) speeds for typical motors with different pole numbers and supply frequencies

No of poles (p)	Synchronous Speed at 50 Hz r/min	Synchronous Speed at 60 Hz r/min
2	3000	3600
4	1500	1800
6	1000	1200
8	750	900
10	600	720
12	500	600

There are two basic sets of standards, one issued by the IEC and defining metric dimensioned, kW rated motors. These are "Harmonised" within the EU by CENELEC and become "EN" standards.[1] NEMA has developed the other set of standards in North America, which define inch/horsepower designs.[2]

Neither basic standard specifies the efficiency or power factor of the motor in question. However, in order to achieve the physical dimensional constraints laid down, there is a practical minimum efficiency inherent in each standard. This is relatively low and a wide range of efficiency and other characteristics can be offered. Within the EU efficiency bands have been agreed. In North America, minimum efficiency standards are mandated by the EPACT legislation, for more information refer to Appendix A1.

No motor is 100% efficient, and therefore its design must take into account the removal of the losses, which are predominantly in the form of heat. Heat from the stator can be dissipated through the motor carcass, however heat from the rotor is generally passed down the shaft, and can increase the bearing heating.[3]

As the windings of motors generate losses, the insulation materials used will also have to cope with the resulting temperature rise. The specifications for winding insulation are based on the temperature rise capacity of the materials. In general the specification is based on a 40°C ambient, with Class B materials being capable of a rise of 80 Kelvin, Class F of 105 Kelvin and Class H of 125 Kelvin for continuous duty (NEMA service factor 1.0). Motor outputs are quoted on the basis of a specific temperature rise i.e. 75 kW with Class B rise, while the motor may well be wound with Class F materials. Class H insulation is more costly and less usual in AC motors.

[1] IEC/EN 60034 – 1 onwards, define metric motors.
[2] NEMA MG1 Standards define inch/hp motors.
[3] See IEC 60034 Part 17 and Part 25 plus the NEMA Guide to Variable Speed Drives.

6.2.3 Output torque

It is conventional to designate motor output in terms of power, but it must be remembered that the motor is in fact a source of torque.

For a variable speed system it is always necessary to consider the required torque/speed characteristic. Many motors, both fixed and variable speed, are designed specifically to match the load speed/torque demand.

The physical size of a motor is directly related to the torque it can deliver.

Catalogue values for two typical IEC motor frames fed at 50 Hz and for two typical NEMA motor frames fed at 60 Hz, are shown in Table 6.2. This shows how similar torque levels are generated by the frame, irrespective of the number of poles.

Table 6.2: Motor frame output relationship for two typical motors

IEC Motor Frame		**160L**		**280M**	
Number of Poles	Synch. Speed	Power [kW]	Torque [Nm]	Power [kW]	Torque [Nm]
4	1500	15	96	90	580
6	1000	11	109	55	530
8	750	7.5	100	45	580
NEMA Motor Frame		**286T**		**445T**	
Number of Poles	Synch. Speed	Power [hp]	Torque [ft-lb]	Power [hp]	Torque [ft-lb]
4	1800	30	132	200	589
6	1200	20	132	125	552
8	900	15	132	100	591

6.2.4 Induction motors operated at variable speed

Induction motors, which are the most widely used drivers for pumps in the industrial and in the residential market sectors, can in most cases be operated at variable speed, by an electronic frequency converter, without modification.

By applying the appropriate variable frequency and variable voltage waveforms to the stator windings, the motor speed will vary proportionally to the applied frequency.

Some specific guidelines do apply to selecting a suitable motor, or to retrofitting a drive to an existing motor (see Chapters 10 and 11).

6.3 Alternative electrical designs of motors

6.3.1 Synchronous motors

6.3.1.1 Separately exited

For high powers, typically greater than 5000 kW (7000 hp), for some high speed and also for some special low power applications, the most usual solution is a synchronous motor, which has relatively lower rotor losses, and is therefore simpler to keep cool.

Synchronous motors may be fitted with a permanent magnet generator to provide the rotor field or a wound exciter on the rotor. With a synchronous motor there is no slip.

6.3.1.2 Switched reluctance motors

Switched reluctance motors are both simple and robust in construction.

The rotor of a switched reluctance motor comprises electrical steel laminations mounted on the motor shaft. Unlike most technologies, the switched reluctance motor has no rotor windings or permanent magnets. The motor is self-excited and uses a transversely laminated rotor with a special geometry to generate a suitable field.

The stator is similarly straightforward, comprising a core of laminated electrical steel and a number of independent phase windings.

The operation of the switched reluctance motor is best described in conjunction with Figure 6.5.

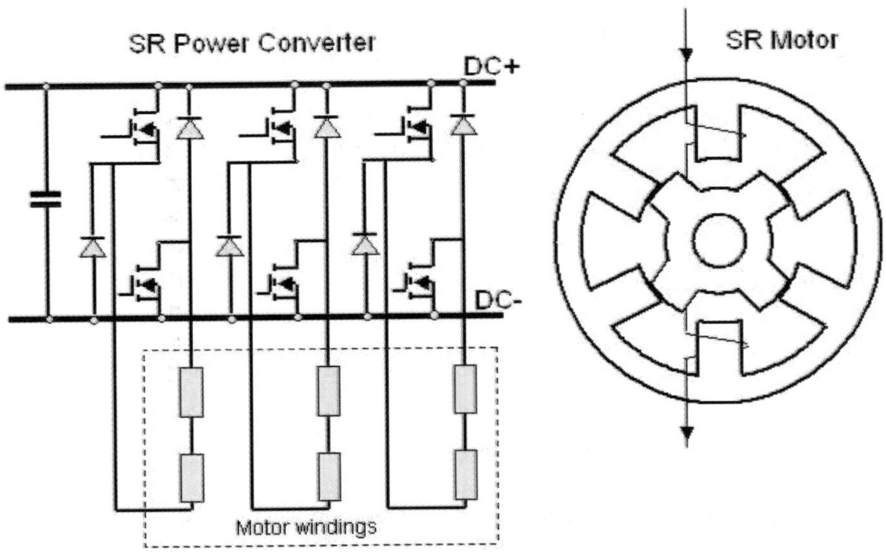

Figure 6.5: Schematic representation of switched reluctance power converter and motor

Diagonally opposed poles are energized, and due to the current flowing, torque is produced in the associated rotor poles. As the rotor poles align, the converter applies energisation to the next pair of poles, to maintain rotation. Most systems require a simple position sensor attached to the rotor to enable the converter to switch correctly.

Switched reluctance motors must be operated in conjunction with a suitable drive unit. This ensures their suitability for variable speed applications, but may limit their application for lower power fixed-speed requirements where the cost of the controller may not be justifiable.

6.3.1.3 Permanent magnet (PM) motors

In recent years, more and more emphasis has been placed on overall high-efficiency characteristics. Due to this, permanent magnet motors are being increasingly considered for many applications.

AC permanent magnet motors generally have stator windings similar to induction motors but they use permanent magnets in the rotor instead of a squirrel cage. DC permanent magnet motors have two or four windings and use electronic commutation to achieve the rotating stator field. The permanent magnet rotor synchronously follows the rotating field in the stator and responds to varying frequencies of the field in a way which can be considered quite similar to the induction motor.

Since the magnet requires no electrical power, unlike an induction machine rotor, the permanent magnet motor is inherently efficient especially at reduced speed and partial loads. The efficiency of a permanent magnet motor tends to be higher than those for an asynchronous motor of similar rating.

Permanent magnet motors have traditionally been used in large numbers for specialised applications, such as spinning machines and machine tools. Due to the requirement of energy saving, high efficiency has become a major selling point in many areas, such as domestic and OEM applications including air conditioners, fans and pumps. However, this technology is maturing and it is increasingly being applied in industrial applications, including pumping. PM motors are limited only by the peripheral speed of the rotor, and are available for a wide range of small power high-speed applications and increasingly in higher power slow speed designs.

Some permanent magnet motor designs require a variable speed drive in order to operate, which has the benefit of enabling energy saving speed control.

The use of a variable speed drive does tend to increase the initial installation cost, which may be difficult to justify in applications where variable speed is not required. However, recent developments in 'line-start' permanent magnet motors (which can operate directly from the network mains supply) enable the efficiency benefits of permanent magnet motors to be realised without the cost penalty of a variable speed drive.

6.3.1.4 The principles of electronically commutated permanent magnet motors

Figure 6.6 shows simplified, the appropriate switch-on phases (current flow) of 4 stator coils. Due to the current flowing in the stator pole windings, a magnetic stator field with north and south pole is formed depending upon the relevant current direction. The permanent magnet rotor, with a constant magnetic field rotating in the stator field, is made to rotate by the attraction and rejection forces of unequal and equal poles. The uniform rotation results from the current flow of the individual stator coils, which circulates with time.

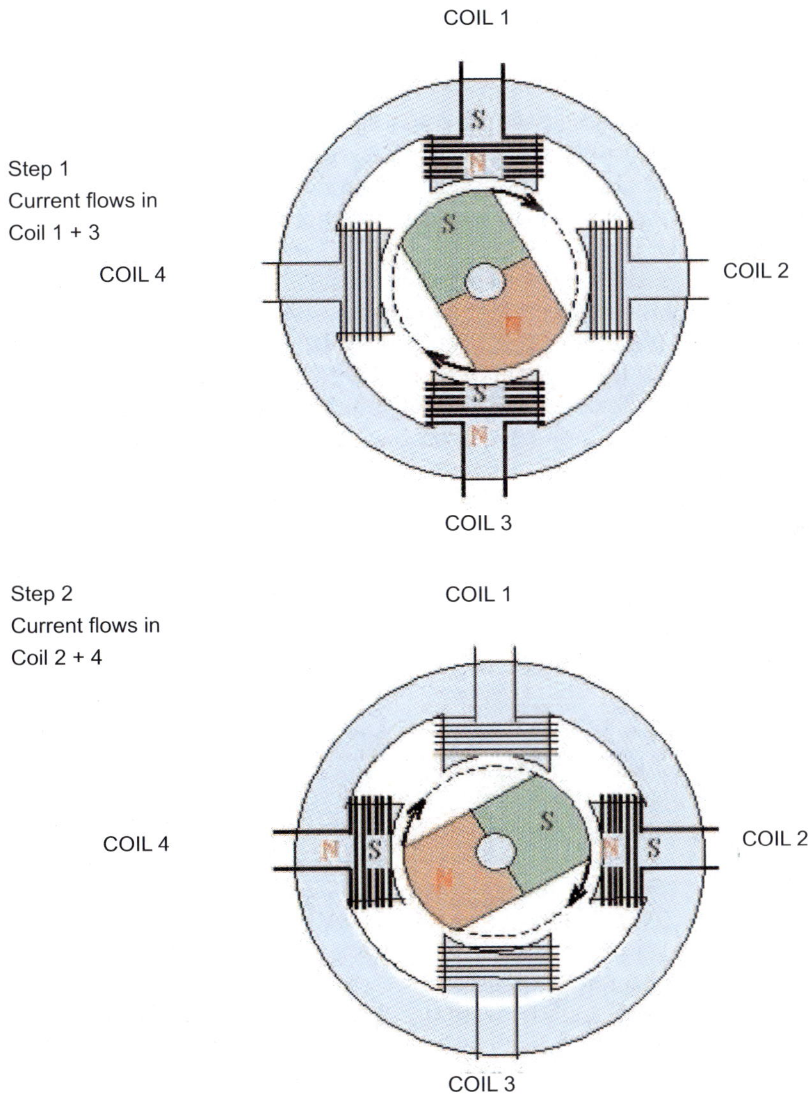

Figure 6.6: Switch-on phases of 4 stator coils

WARNING – Great care needs to be exercised with PM motors in applications where energy can be taken from the driven load, as they also operate as permanently excited generators. This means that back flow in pumping installations, which causes reverse rotation, must be avoided when permanent magnet motors are used.

6.3.2 DC motors

The conventional DC motor has a wound stator to provide a field. A wound rotor (the armature) is fed by a commutator, which switches the power to the appropriate windings to give an offset magnetic field to develop the torque.

Traditionally, the lower cost, greater simplicity, and ease of control of DC drives compared to AC drives, was much more significant than the higher cost and lower efficiency of DC motors. DC drives were therefore the norm for speed control applications for many years. However, the higher maintenance requirements and generally lower environmental protection of DC motors are often held against this technology.

DC drives are still used where some advantages are perceived to exist, but even these are being eroded as AC drive technology improves. Developments such as flux vector control have now allowed AC drives to be used in high performance and highly dynamic applications where accurate torque/speed control is required. This used to be the exclusive benefit of DC drives.

DC permanent magnet motors are described in Section 6.3.1.3.

6.4 Motor construction and cooling

All electric motors have losses, mainly manifested as heat, which must be dissipated. This is normally to the ambient air, and a conventional motor is designed to dissipate this heat when running at fixed speed and full output.

6.4.1 Dry installed motors

The majority of electric motors are designed to operate with ambient air crossing the motor frame to achieve cooling. Fans mounted directly to the motor shaft generally achieve the airflow. In special cases, especially with large power motors, a heat exchanger, either air/air or water/air, may be used.

The levels of ingress protection for the motor and the various cooling forms are defined in the IEC and NEMA standards.

6.4.2 Submersible pump motors

In principle, most submersible pumps use squirrel-cage induction motors that are not significantly different from the typical non-submersible induction motors, except with regard to cooling.

Submersible pump motors, which are not provided with internal or external cooling systems, are cooled by full or partial submergence in the pumped liquid. In such cases motor cooling is not normally affected by speed reduction when driven by a VFD.

Submersible pump motors, above a certain size, are provided with various types of cooling systems to dissipate the heat generated from motor losses. Some of these designs have minimum speed requirements that must be met to assure proper cooling.

For installations which can be either dry or immersed, a submersible pump motor is normally provided with a cooling jacket around the stator casing, inside which a cooling liquid is circulated.

Submersible borehole pump motors must be installed such that the flow of the pumped liquid provides sufficient motor cooling. This is often achieved by the use of a shroud if units are used in non-confined areas.

6.4.3 Wet rotor motors (canned motors)

In other applications a “wet rotor” or “canned motor” construction is utilised, with a physical non-magnetic barrier between the stator and rotor of the machine. The use of a canned design eliminates the requirement for seals around rotating elements.

As it is relatively insensitive to the gap between the stator and rotor, permanent magnet motors tend to be more and more widely used in small power water pumps, up to 4 kW (5 hp) when the preferred solution is the so-called “wet rotor” configuration.

The efficiency benefits of permanent magnet motors also apply to “wet” rotor motors. Table 6.3 shows a comparison of efficiencies for an asynchronous “wet” induction motor and a “wet” permanent magnet synchronous motor, of equal size and gap.

Table 6.3: Comparison of motor efficiencies

TYPE OF MOTOR	INPUT POWER KW	OUTPUT POWER kW (hp)	EFFICIENCY %
“WET”, INDUCTION	4	2.9 (3.9)	72.5
“WET” PERMANENT MAGNET	3.3	2.9 (3.9)	88

Other special designs include motors, which are totally liquid filled (water or oil), or allow free ingress of the pumped medium.

6.5 Motor starting

Fixed speed motors will normally be connected direct on line (DOL) to the electrical supply network, and will cause a high starting current to flow on switching. The magnitude of this current may be problematical with larger power motors or weak power supplies. There are several alternative means of providing control of this starting current, which can avoid network problems. Typical starting characteristics are shown in Figure 6.7.

Without control, most motors will have a limited number of starting cycles per hour, due to the slip energy causing a temperature rise in the rotor during starting.

Variable speed drives eliminate high starting currents and rotor heating on starting and hence there are generally no restrictions on the number of starts. It is also possible to reduce the power rating of standby generating equipment because of the lower peak current. Lower peak currents can also reduce peak demand charges.

Currently employed starting methods include "reduced voltage" starters such as star-delta (Wye-delta), autotransformer or electronic soft starters, however they all result in reduced torque availability at breakaway. In general this is not a problem with rotodynamic pumps, but can be problematic with some PD pump types.

Formerly one of the most common solutions was to use a wound rotor motor, which has rotor resistance starting. However, the higher life cycle costs (initial cost, maintenance cost and reduced efficiency) of this solution has caused its demise.

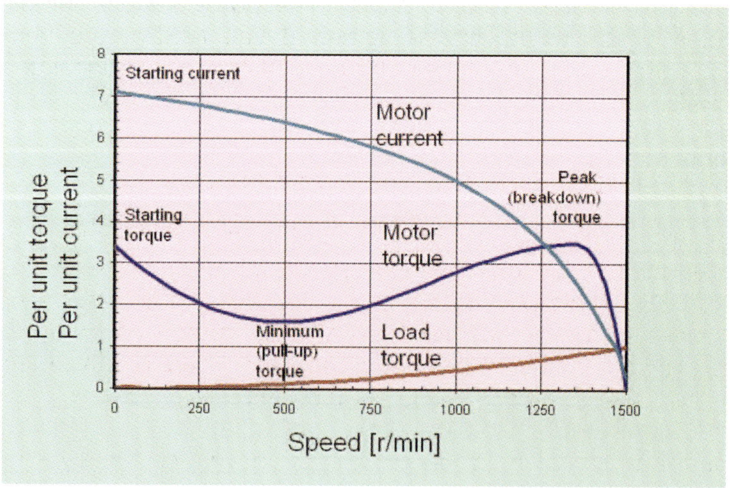

Figure 6.7: Induction motor DOL starting characteristics

CHAPTER • SEVEN

Variable speed drives

7.1 Types of variable speed drive

A large number of types of drive are offered for the motor types previously described. Some types of drive, especially electro-mechanical controllers ,are now considered to be outdated technology, but are included for completeness (see Figure 7.1). When a motor and a drive are combined, this becomes a power drive system (PDS). The power ranges for different technologies of electrical variable speed drives (VSDs) are shown in Figure 7.2, with a summary of their characteristics and additional details in Appendix A2.

7.2 Variable speed drives for induction motors

There are two ways of varying the speed of an induction motor, either by varying the motor slip or by varying the supply frequency.

Whilst it is possible to vary the slip of a conventional cage induction motor by reducing the supply voltage, this introduces unacceptable additional losses manifested as heat in the rotor. Therefore this method is only practical for wound rotor motors where this energy can be dissipated externally.

The preferred practice for electrical speed variation is by changing supply frequency with a variable frequency drive (VFD). The most common form of VFD system is the voltage source pulse width modulation (PWM) frequency converter (often incorrectly referred to as an inverter) applied to a cage induction motor.

Many other designs have been developed. However, few are still offered except for specialised applications.

The frequency converter produces a variable frequency, variable voltage, output waveform. In its simplest form the converter develops a

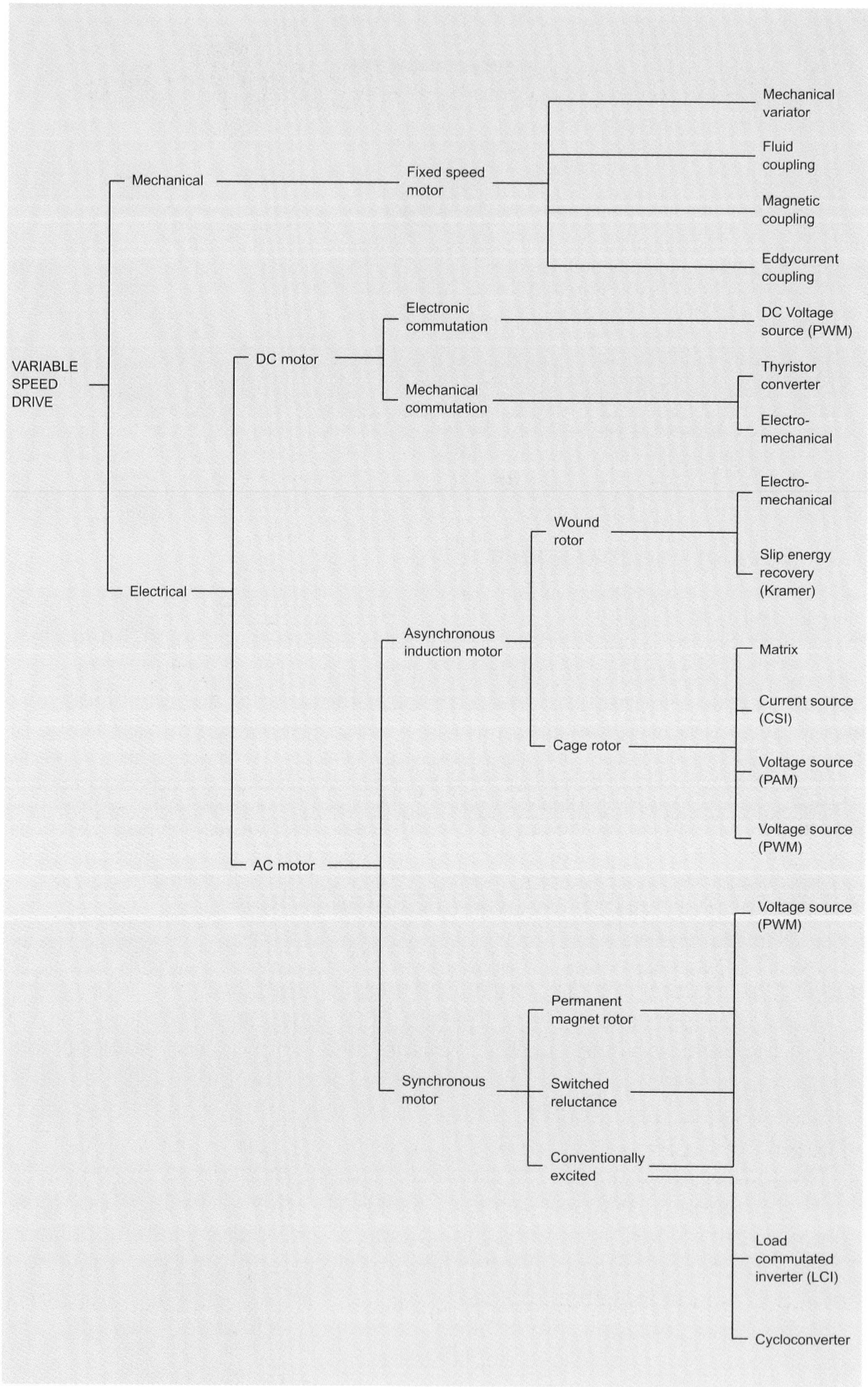

Figure 7.1: Types of variable speed drive (VSD)

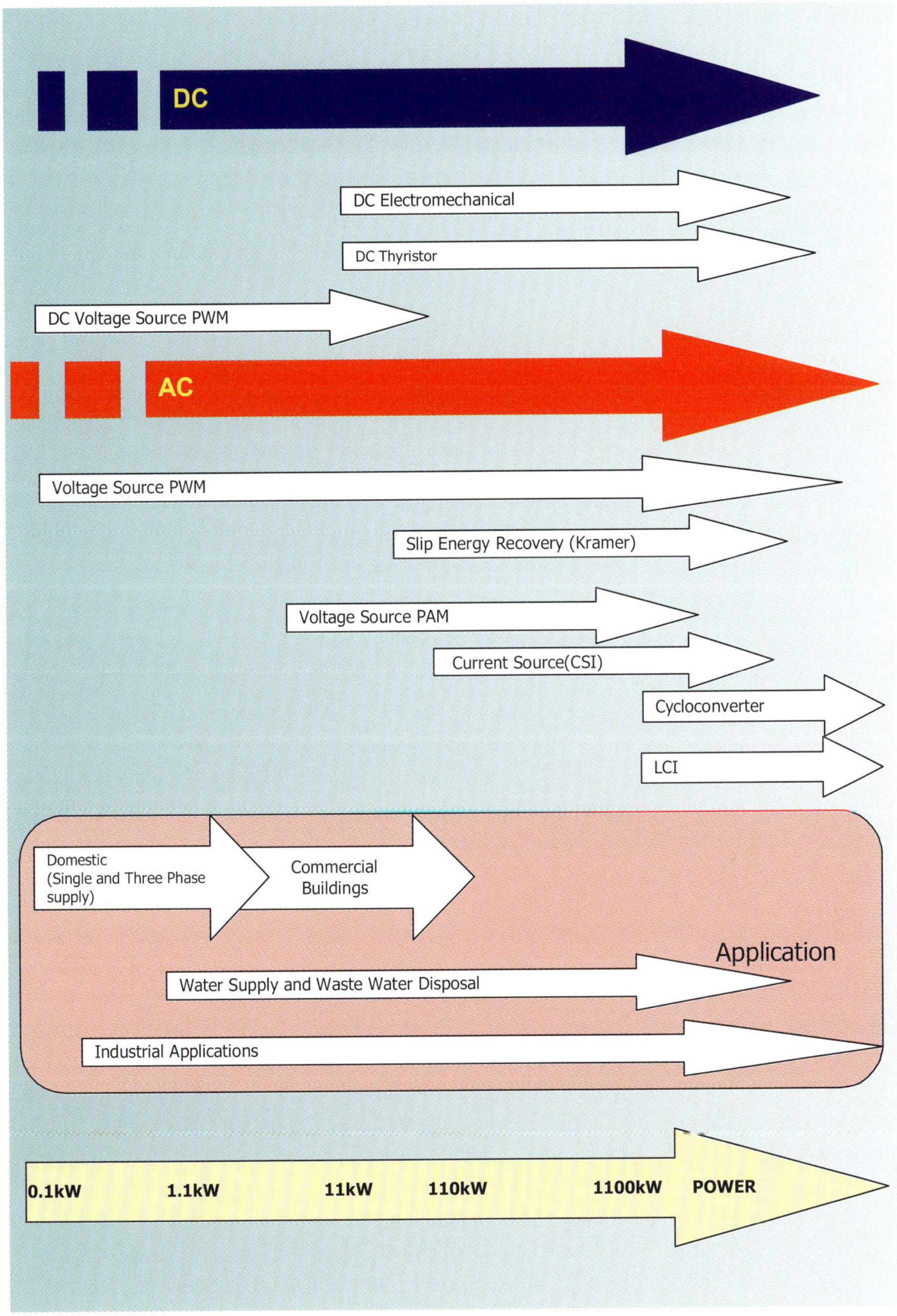

Figure 7.2: Power ranges and application fields for electrical VSDs

voltage directly proportional to the frequency, which produces a constant magnetic flux in the motor. If such waveforms are applied to the stator windings there will be a new discrete speed torque relationship for every discrete frequency that can be generated. The motor speed will be proportional to the frequency generated by the VFD. Figure 7.3 illustrates the effect on a four pole motor operating at three discrete frequencies and showing a typical rotodynamic pump speed torque relationship and a typical positive displacement pump speed torque relationship.

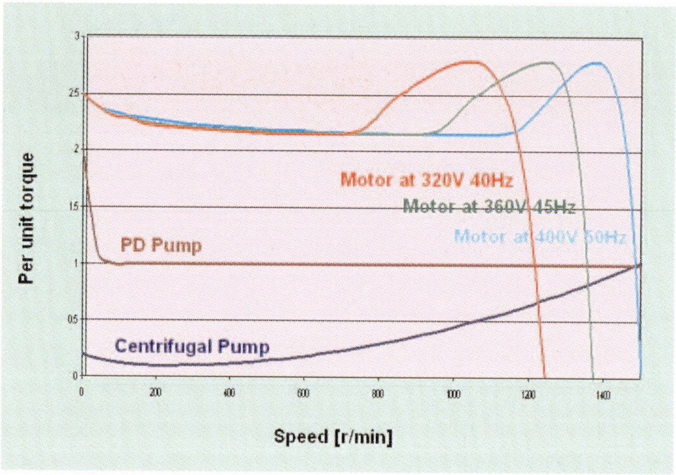

Figure 7.3: Speed/torque curves for a four pole cage induction motor when fed at three discrete frequencies at constant magnetic flux

In theory a motor can be driven at any speed between zero and its design (base) speed by maintaining a constant flux. If a higher speed is required, this may be accommodated, providing the motor is appropriately designed. However, as the inverter output voltage will be limited, the motor flux and consequently the available torque will be reduced.

Figure 7.4 shows the design of most frequency converter drive systems. The three-phase AC voltage is initially rectified to a direct current (DC), then smoothed and finally, the DC/AC inverter converts the DC voltage to the variable voltage and variable frequency output applied to the motor. Further details of drive design are described in Appendix A2.1 and A2.2

The pulsed output voltage is applied to the motor and the resulting current, modified by the significant motor inductance, consists mainly of the fundamental sinewave at the required operating frequency with a superimposed low magnitude ripple component based on the switching frequency. Both voltage and current over one cycle are illustrated in simplified form with deliberately reduced switching frequency in Figure 7.5. The voltage and current waveforms are not in phase due to the effects of the motor power factor.

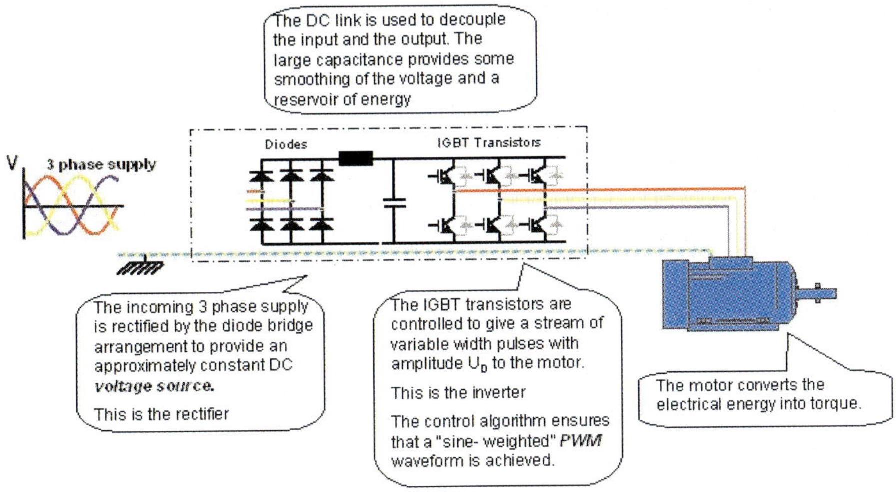

Figure 7.4: Conventional frequency converter design [1]

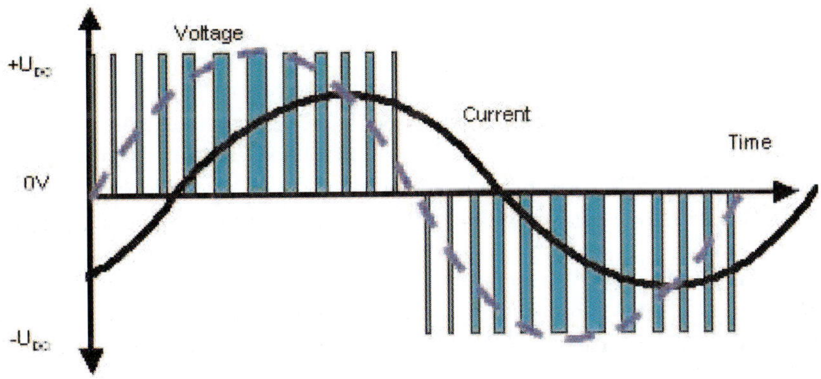

Figure 7.5: Simplified PWM frequency converter output

7.3 Variable speed drives for other AC motors

In general the voltage source PWM concept is also applicable to the majority of alternative motor designs. This principle is equally applicable in basic terms to the conventional and permanent magnet synchronous motor, and with some small adaptation to the switched reluctance motor. However, an exciter control is also required for a conventional synchronous motor.

A summary of possible alternative drive and motor combinations is shown in Table 7.1.

[1] Reproduced from REMA/Gambica Technical Report No1

Table 7.1: Summary of drive applicability

Converter	Asynchronous Motor	Synchronous Motor
Voltage source PWM	Applicable	Applicable
Voltage source PAM	Applicable	
Current source inverter (CSI)	Applicable	
Load commutated inverter (LCI)		Applicable
Slip energy recovery (Kramer)	Applicable to wound rotor machines only	
Cycloconverter	Applicable	Applicable
Matrix	Applicable	
Switched reluctance drive (SRD)		Special form of voltage source PWM converter applicable to switched reluctance motors

7.4 Variable speed drives for DC motors

A DC commutator motor requires a variable current to be applied to the armature, which is normally obtained by a controlled rectifier bridge, whereas the AC drive design illustrated in Figure 7.4 uses an uncontrolled rectifier. The DC motor drive also requires a field supply, which may be a constant voltage, or a controlled voltage in more complex control schemes.

Brushless DC motors use a modified PWM converter.

7.5 Drive controller algorithms

Manufacturers often use complex technical terminology to specify the type of control that their specific drive offers. The power electronics components in low voltage PWM drives do not differ very much from one manufacturer to another. An important consideration in the capability of a drive is its ability to deliver and control torque and speed, particularly at lower frequencies.

The control strategies are achieved by different algorithms programmed into the microprocessor of the drive, which ultimately determine the output torque of the motor. The algorithms require information about the motor in order to be effective. The characteristics of the motor such as volts, full load current and base speed are entered in the drive configuration during set up. Other parameters require the drive to perform a tuning routine on the motor to measure static and dynamic parameters. The drive monitors volts, amps and frequency continuously when running and may also measure speed and position of the rotor if an encoder is mounted on the motor shaft and fed back to the drive. The drive uses the motor information to model its response and to control the firing of the power transistors with the objective of obtaining optimum torque and control.

For centrifugal pumps it is unlikely that any special form of control is required as they are low starting torque loads. Special control forms may give secondary benefits from better motor control. With a positive displacement pump much more care is required, as the power drawn is roughly proportional to the rotational speed, and in addition a high starting torque may be required especially for progressive cavity pumps (refer to Appendix A2.3 for further details).

Using the spare capacity of the drive microprocessor can allow control optimization to be built into a package.

7.6 Energy optimizing or "flux optimizing" techniques

Many VFDs incorporate an "energy optimizing" feature, which saves energy by reducing the motor magnetizing current at low mechanical loads. The principle is that at low load, with a centrifugal load, the motor does not need to produce such a strong magnetic field to produce torque. These efficiency improvements are relatively small; typically at 50% load and 90% speed, for a 22 kW (30 hp) motor, the inverter fed efficiency may be raised from 92% to 92.5%. This is usually much less than from the slowing down of the driven machinery, but still gives a tangible benefit.

Energy optimizing will also result in an improvement in the motor noise and vibration.

7.7 Power drive systems

In all types of rotating machinery, the use of a VFD frees the designer from the traditional constraints of designing machinery for the specific rotational speeds detailed in Table 6.1, and even allows operation at higher speed. In many instances, pump designers have to develop products to be sold in markets with both 50 Hz and 60 Hz base frequencies and best efficiency may be achieved at some point between these levels. Particularly with centrifugal equipment such as fans or pumps, VSDs allow machinery to be operated closer to optimum specific speed for a much wider range of duties.

Traditionally VFDs have often been applied to existing and new motor systems as a stand alone component, placed between the power network and the motor. As the VFD market is maturing the number of products continues to increase. Packaging of motors and drives, as well as the concept of integrated solutions and integrated motor/VFD units, are now readily available as alternatives to purchasing stand-alone drives.

When a stand-alone drive is applied, it is important to ascertain that it is compatible with the motor, the motor cabling and the power network system. Drive motor systems available from system integrators, as well as packaged systems generally assure that the selected components will work well together. The integrated motor/VFD unit, described in more detail below, represents a virtual guarantee for full compatibility between drive and motor.

7.8 Integrated motors and drives

A number of motor and drive manufacturers are now producing integrated motor/VFD units (see examples in Figure 7.6). These units consist of a motor and a specially designed VFD, produced as a single package, with the VFD unit mounted variously on the top, side, or end of the motor. Currently these designs are available up to approximately 22 kW (30 hp).

Integrated motors and drives can feature a number of advantages:

- Motor and VFD are properly matched
- Installed costs can be less than that of a separate motor and VFD
- No special cables between VFD and motor, saving installation time and screened cables
- Reduced over-voltage on the motor windings due to the short length of the connections, leading to a longer life of the insulation
- No output cables. Hence radiated emissions are minimized
- VFD shares motor cooling system
- Better compatibility between all the components of the drive system, reduces noise & vibration
- Often easier and faster to retrofit in a building or industrial installation

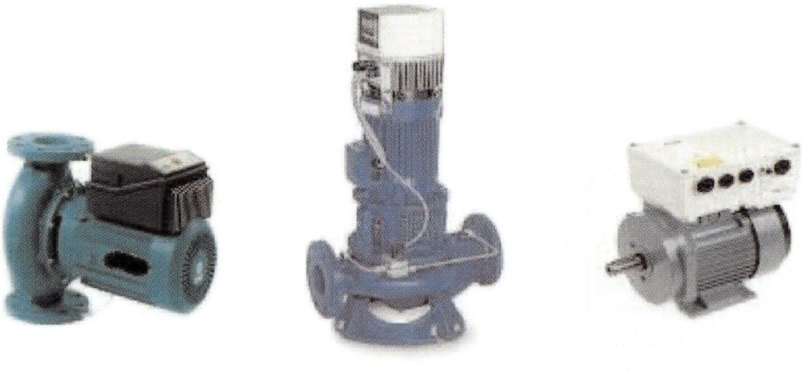

Figure 7.6: Illustrations of integrated motor & VFD

CHAPTER • EIGHT

Control principles for variable speed pumping

8.1 Methods of controlling a rotodynamic pump

The methods of controlling a pump for varying duty are covered in more detail in other sections, but as an introduction to control principles a summary is presented.

When the user wants to reduce the flow in a process, valve control (see Figure 8.1) can be used either to directly throttle, or control a bypass (see Figure 8.2). Alternatively, speed control (see Figure 8.3) can be applied using a VSD.

Valve control will effectively reduce the flow, however, the consumed energy is often significantly lower if speed control is used. On-off control can be used to vary pumped volume in systems where an intermittent flow

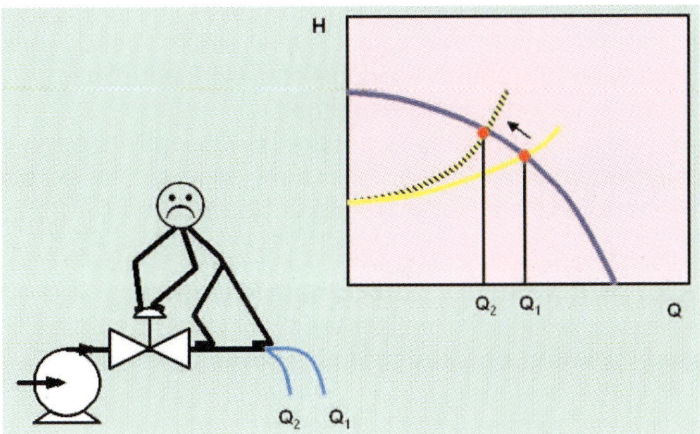

Figure 8.1: Throttle valve control

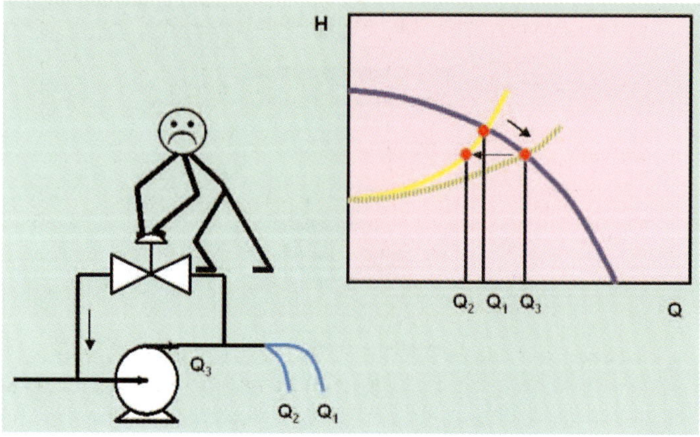

Figure 8.2: By pass control

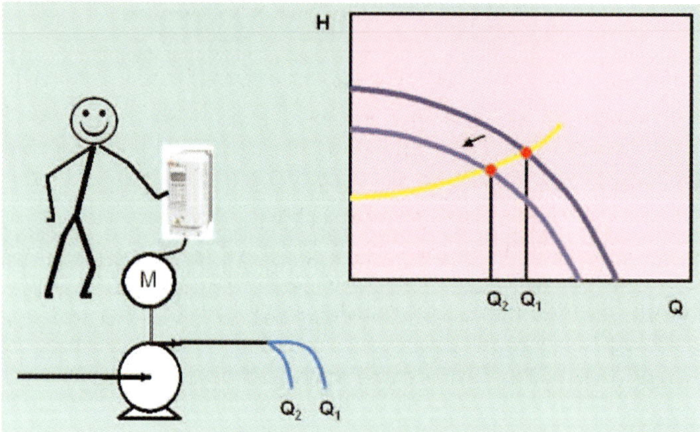

Figure 8.3: Speed control

is acceptable. These systems often require a storage facility.

If there is a static head in the pumping system, a substantial amount of energy may be used for lifting, compared with the energy required to overcome the friction in the pipes. Only small speed variations are possible and although energy savings can still sometimes be achieved, applications must be individually evaluated.

In many pumping applications several pumps are used in parallel to produce the required flow when the flow demand varies substantially (typically water and waste water systems). See Section 4.2.2.3.

8.2 Control principles for rotodynamic pumps

8.2.1 Speed variation of rotodynamic pumps

In applications that require flow or pressure control, particularly in systems with high friction loss, the most "energy effective" technique is to use

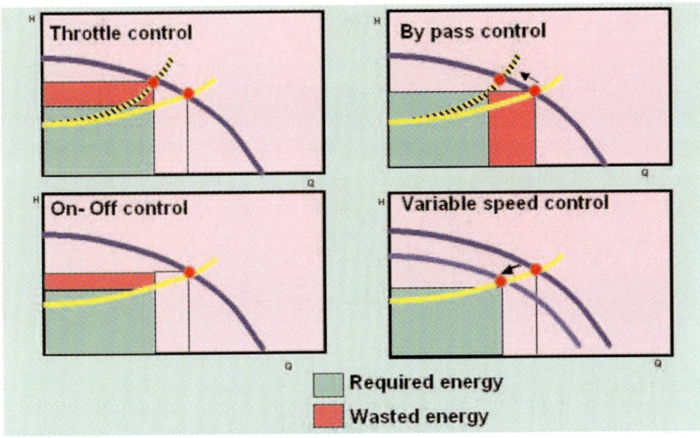

Figure 8.4: Energy consumption with constant pump and motor efficiency

electronic variable frequency drives (VFDs), which can match the motor speed to the load requirement (see Figure 8.4). This eliminates a number of costly and energy inefficient ancillaries such as throttle valves or by-pass systems. At the same time, VFDs normally provide close motor control during start-up of the pump. The current that flows into a PWM drive rectifier is proportional to the power being delivered, plus losses. Therefore the drive also gives a very effective soft start to the power supply network, as well as a controlled start of the motor.

8.2.2 Control at constant pressure with varying flow rate

a) Single pump system

The most common form of control is by use of a discharge pressure sensor, which sends a signal to the VFD. This in turn varies the speed allowing the pump to increase or decrease the flow required by the system

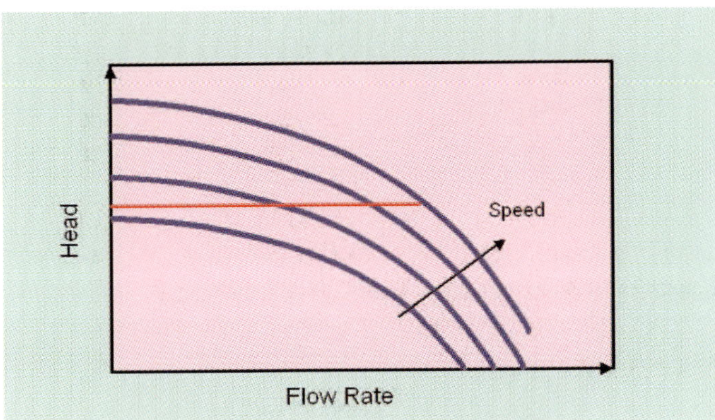

Figure 8.5: Constant pressure

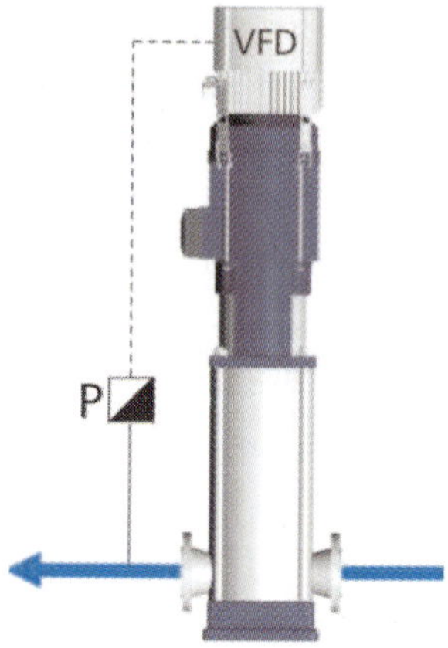

Figure 8.6: Typical VFD pump for water supply

while holding the pressure constant. See Figure 8.5 for the pump curves at different speeds and Figure 8.6 for the pump arrangement with pressure sensor. This form of control is common in water supply schemes where a constant pressure is required but water is required at different flow rates depending on the number of users at any given time. Also, capacity changes at constant pressure are common in centralized cooling and distribution systems and also in irrigation systems where a varying number of spray heads or irrigation sections is involved.

b) Multipump system

If the flow variation for most of the working time is substantial, such as 5 to 1 or as little as 2 to 1 and the system curve is flat, it may make sense to use more than one pump to reach maximum flow. This will improve the efficiency at low demand (see Figure 8.7).

A sequence control system will start up the pumps depending on the actual demand and will also do automatic, cyclic changeover of the pumps

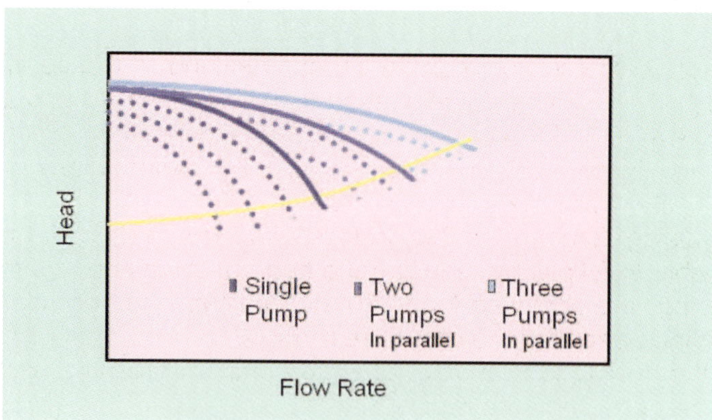

Figure 8.7: Multipump curve

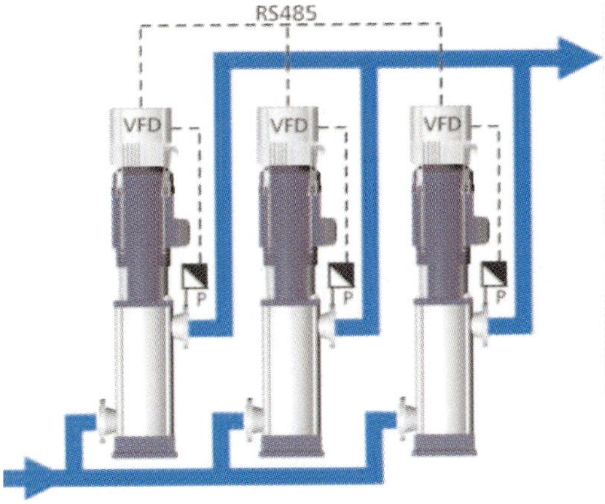

Figure 8.8: Booster application

to get equal running times. Intelligent systems can also increase the output pressure depending on the demand to compensate for friction losses in the pipework.

Multipump systems can utilize a single VFD to control one pump in combination with a number of fixed speed (slave) pumps, however careful economic and hydraulic analysis is required.

Important: in pressure controlled supply systems, the pumps must shut off immediately when the demand stops. For booster pump sets (illustrated in Figure 8.8), this must happen independently of varying incoming pressure and therefore with varying pump head at zero flow.

Operation of pumps at zero flowrate wastes energy but more seriously this causes a high temperature rise within the pump and must therefore be avoided.

8.2.3 Control by differential pressure according to a demand curve

This form of control is common for pumps working in closed circulation systems (HVAC). In this application the pump should work along a curve A to B in Figure 8.9, in order to compensate for friction losses in the pipework. Intelligent VFD products can do this without any additional components like a flowmeter or a controller.

At decreasing demand the friction losses are lowered and if the control system is able to follow, by reducing the head of the pump, the maximum energy savings are achieved (up to 70%). This happens because both the flow Q and the head H of the pump are reduced.

If only the flow Q is reduced, like in a constant pressure system, the possible energy savings are reduced.

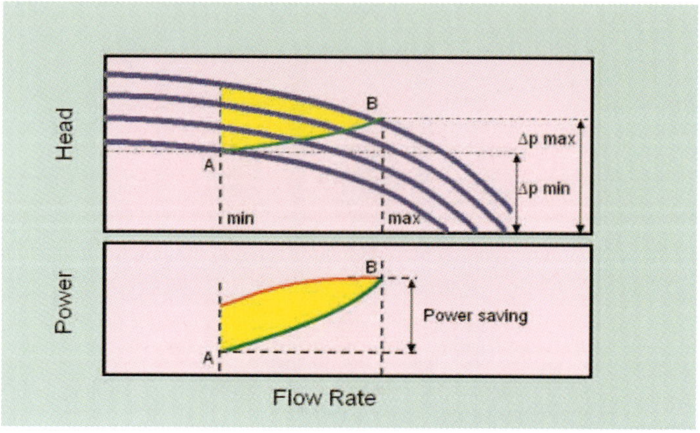

Figure 8.9: Control according to an increasing curve of Δp

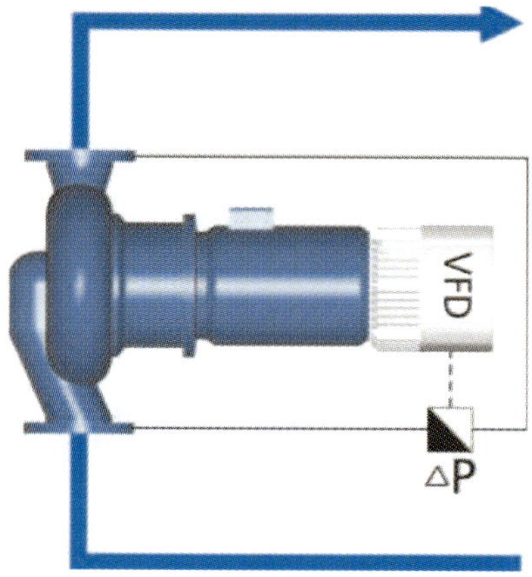

Figure 8.10: Circulation pump with differential pressure sensor

8.2.4 Control by varying pressure at a fixed flow rate

Many cooling, chiller, spraying and washing applications require a specific flow rate to be supplied even if the suction and delivery conditions vary. Typically suction conditions vary when the level in the suction reservoir or tank changes. Delivery pressure will vary when filters start to block, thereby increasing system resistance.

The VFD system is usually the optimum choice to keep the flow rate constant in such a system using a control signal from a flowmeter. The speed increases or decreases and restores the flow to the required value. Figure 8.11 shows constant flow with varying pressure/speed. The general arrangement is shown in Figure 8.12.

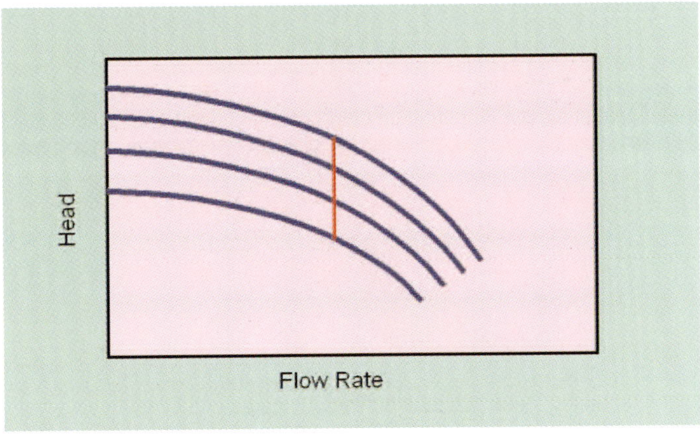

Figure 8.11: Control on constant flow

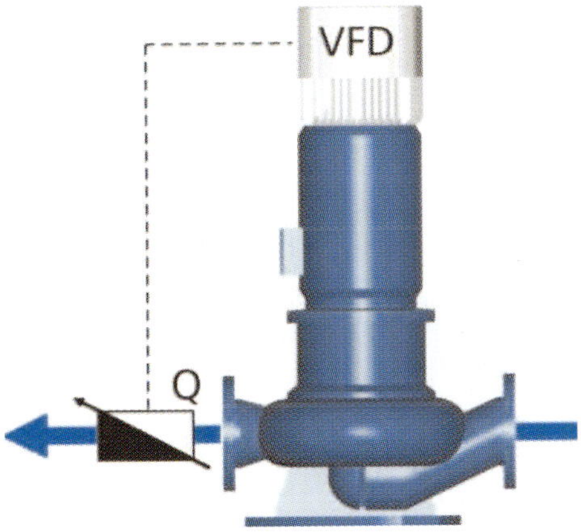

Figure 8.12: Constant flow measurement by flowmeter or orifice plate

8.2.5 Control by varying flow maintaining a fixed liquid level

To maintain a constant fluid level in a tank it is sensible to vary the speed to match the flow of the feed pump to the process demand. An example of this would be an unmodulated steam boiler system. Compared with an on/off controlled pump, this variable speed system not only avoids large level fluctuation in the boiler and in the condensate tank, but also prevents water hammer and a high number of pump starts with full power (Figure 8.13).

This form of variable speed control can also be used successfully on boilers to replace modulation control, eliminating the need for a modulating valve and by-pass line.

A similar method but with reversed function of the control system can be used to pump away the contents of a collecting tank when the inflow

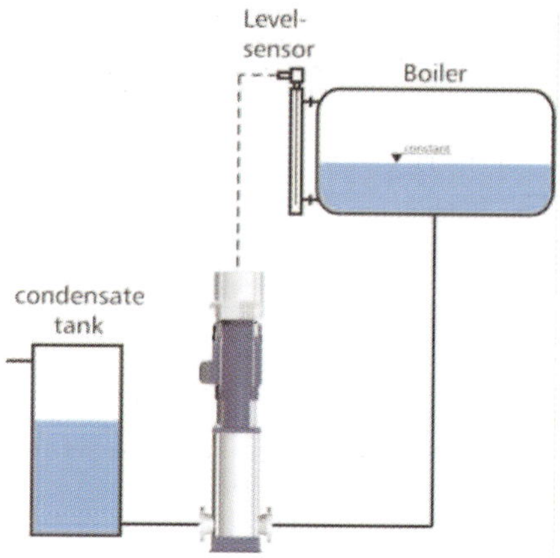

Figure 8.13: Steam boiler feed application

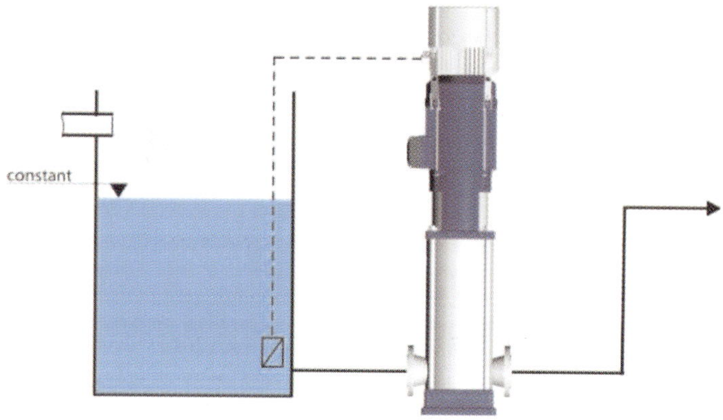

Figure 8.14: Pumping away from a collecting tank

varies. In this way the necessary volume of the tank can be reduced to approximately 10–20% of the volume needed for an on/off controlled fixed speed pump (Figure 8.14). In this application a falling level in the tank, results in a decreasing pump speed (inverse function of the controller).

8.2.6 Control by varying flow maintaining a fixed temperature

In heating and cooling systems there is a requirement for flow to vary based on temperature.

In this instance the VFD is controlled by a temperature sensor which allows the flow of hot or cold liquid in the system to increase or decrease based on the actual temperature required by the process. This is similar in operation to pressure control, where the flow is the variable entity, but a constant temperature requirement from a temperature sensor replaces that from a pressure sensor see Figure 8.15.

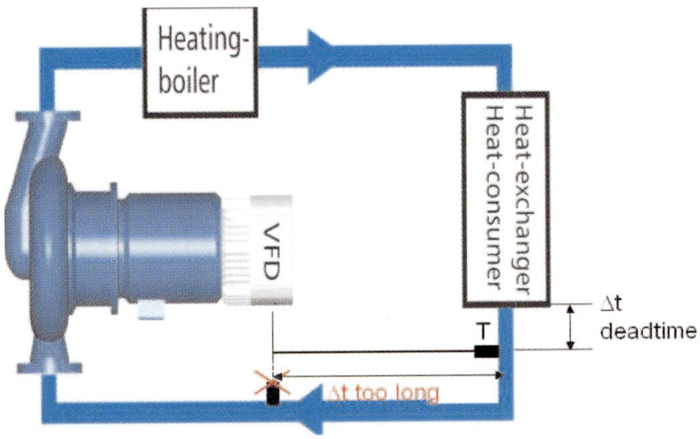

Figure 8.15: Maintaining a fixed temperature in a system
Important: Any dead time (Δt) must be very short

8.2.7 Suction pressure monitoring to avoid cavitation

In cavitation prone applications a suction pressure sensor can be installed close to the pump to reduce pump speed/ capacity, which reduces NPSHR and increases NPSHA, to protect the pump.

8.2.8 Controlling flow by using the power/flow relationship of the pump

A control algorithm allows the pump speed to be varied without measuring external values. It needs no external sensors but a monotonically increasing power curve. The algorithm for this relationship is stored in the VFD microprocessor.

The pump speed is controlled by the power consumption by the motor, measured within the drive, instead of using a signal from an external transducer. Selected operating points on the system curve (A,B,C) relate to power consumption points (D,E,F) for corresponding frequency values see Figure 8.16.

The control monitors the power consumption and adjusts the pump speed (flow rate) according to the preset value programmed into the algorithm.

8.3 Speed variation for positive displacement pumps

8.3.1 Consideration of starting torque

The speed torque relationships of Positive Displacement (PD) pumps are explained and illustrated in Section 4.3. The starting torque of a PD pump has a major influence on the selection of the motor and the VSD, which

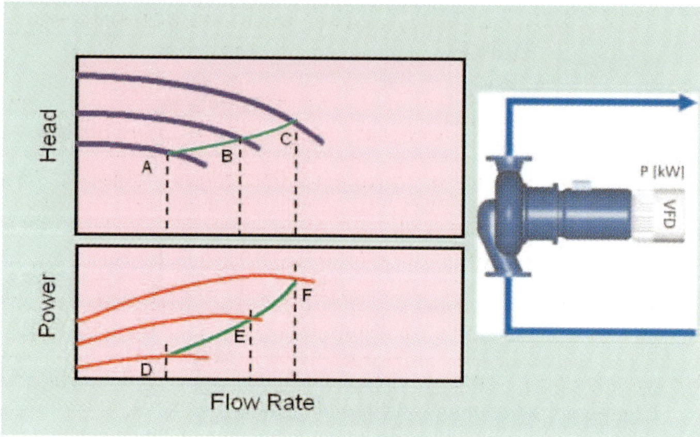

Figure 8.16: Control by pump power algorithm

both have to be suitable to handle the required starting torque. The value of the starting torque is different for each specific design of pump, i.e.

Progressive Cavity	–	(up to) 4 times running torque
Rotary Positive Displacement	–	2 times running torque
Diaphragm (hydraulic)	–	2 times running torque
Piston /Plunger	–	1.5 times running torque

8.4 Control principles for PD pumps

8.4.1 Use of sensors

The VSD controller can use a signal from sensors such as flowmeters, pressure transducers, temperature detectors etc, these sensors must be capable of working at the relief valve accumulation pressure. The accuracy of the signal must be suitable for the application. Many control systems will be similar to rotodynamic applications as described in Section 8.2. However, the major difference is that PD pumps have a direct relationship between speed and flow irrespective of pressure. This makes them particularly suitable for dosing applications.

8.4.2 Control of positive displacement pump by algorithm

An algorithm can be used to control pump speed. The algorithm will have to consider different variables for each specific design of pump and application, see Table 8.1.

Table 8.1: PD pump variables to be considered in speed control algorithms

PUMP TYPE	VARIABLES TO BE CONSIDERED
Progressive Cavity	Stator hysterisis. Viscosity, variable with temperature
Rotary	Volumetric efficiency is varying with viscosity changes and liquid temperature
Diaphragm (Hydraulic)	Volumetric efficiency is varying with viscosity changes to the oil temperature
Piston /Plunger	Volumetric efficiency changes with compressibility

8.5 Implementation of control systems

In many cases there will be an external control system, such as a PLC or a distributed control system (DCS), which will provide the control by sending analogue or digital signals to the drive. In other cases the drive may have adequate on-board intelligence.

All modern drive systems rely on microprocessor control, and this allows the manufacturer to integrate the basic signal processing functions into the drive. In some instances, manufacturers have installed bespoke control software to the drive to allow specific requirements to be met.

8.6 Soft starting and stopping

When an induction motor is started direct-on-line, it generates a high level of torque, which causes a very fast breakaway, and it then accelerates up to speed in an uncontrolled fashion.

The effect on the pump is to place mechanical stresses on the rotating components, followed by stresses in the hydraulic system, which may include a high initial draw-down causing a vacuum to be drawn on the suction side, or a surge on the discharge.

Equally when stopping, the rate of deceleration is totally uncontrolled, which can lead to further mechanical stresses and surges in the hydraulic circuit. This can lead to requirements for additional inertia to be added to a pump, generally in the form of a flywheel, or for surge control vessels in the hydraulic system.

The use of electronic starting systems provides smooth acceleration and deceleration of a drive system.

Electronic soft starters reduce the voltage at the motor terminals in a controlled manner, but are generally short time rated devices, while a frequency converter is usually continuously rated and so can be used to give very controlled rates of change.

The only drawback with either electronic scheme is that the equipment must be connected to the network, and therefore problems could arise in case of a power failure, when an uncontrolled deceleration will occur.

CHAPTER NINE

Selection process – new systems

9.1 Selection for optimum life cycle cost

It is essential to commence the pump system selection with the hydraulic requirements, and to work systematically to select the pump, motor and drive. When the pump maximum duty is known, the peak power and speed for the drive will become clear.

It is common to oversize system components (pumps, motors and drives) however this practice is not recommended since it leads to higher life cycle costs.

When selecting a rotodynamic pump in combination with a VSD, for a system with static head, the pump shall be selected such that the maximum flow rate is to the right hand side of the best efficiency point (BEP). For constant flow regulated systems, the recommendation is to select the pump to the left of BEP at maximum pressure.

All operating conditions must be considered when designing the system (see Figure 5.1). Some operating profiles may best be satisfied by installing multiple pumps, which could be fixed or variable speed. Using the specific energy calculation in Section 5.2 will verify the best solution from an energy point of view. This will also lead to the lowest maintenance costs.

Having established the working range and power required, it is possible to size the motor(s) and drive(s).

9.2 Flow charts

Flow charts to assess the suitability of a VSD for a pump system are shown in Figures 9.1 for rotodynamic pumps and 9.2 for positive displacement

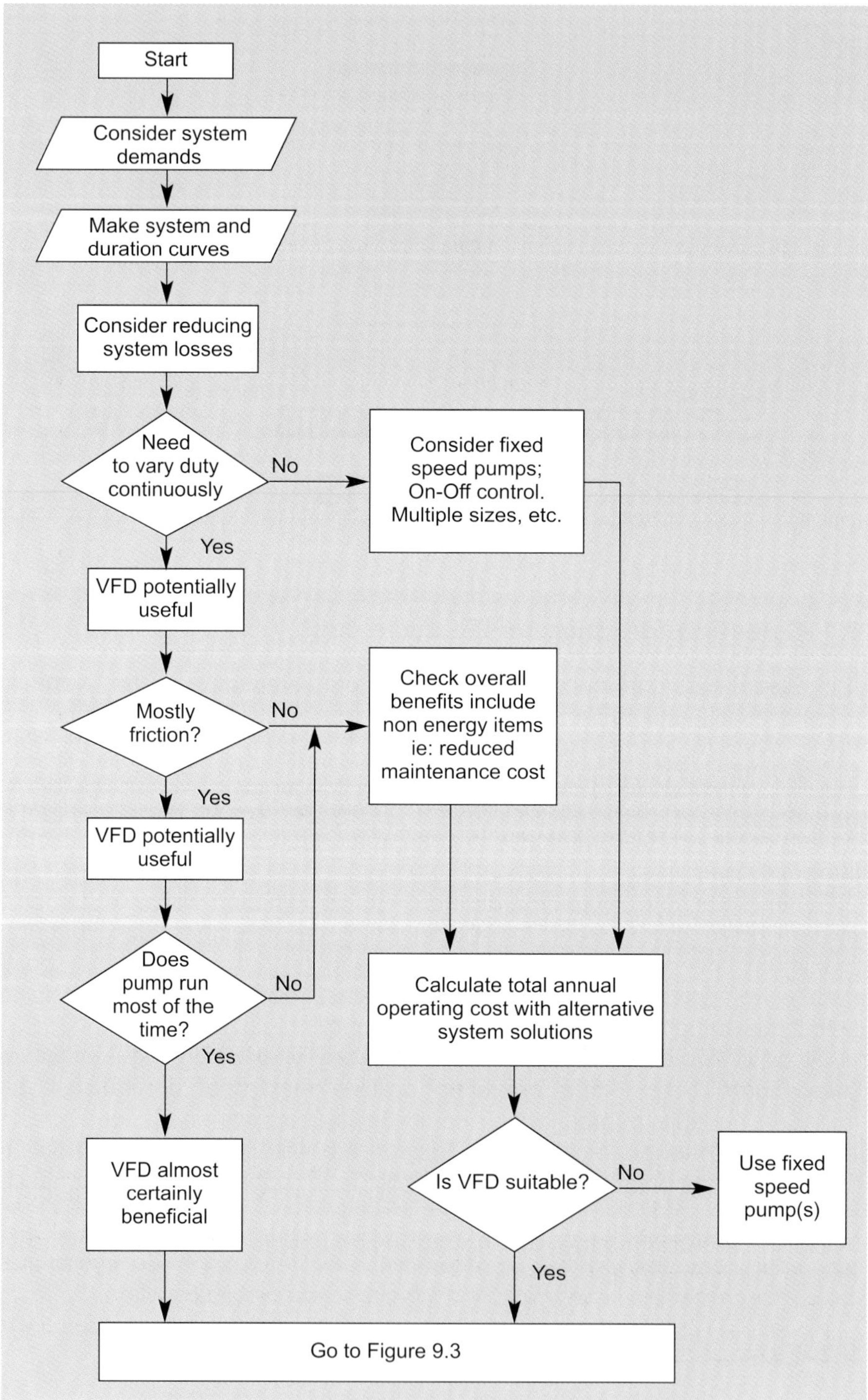

Figure 9.1: Flow chart to assess the suitability of using a VFD in a rotodynamic pump system

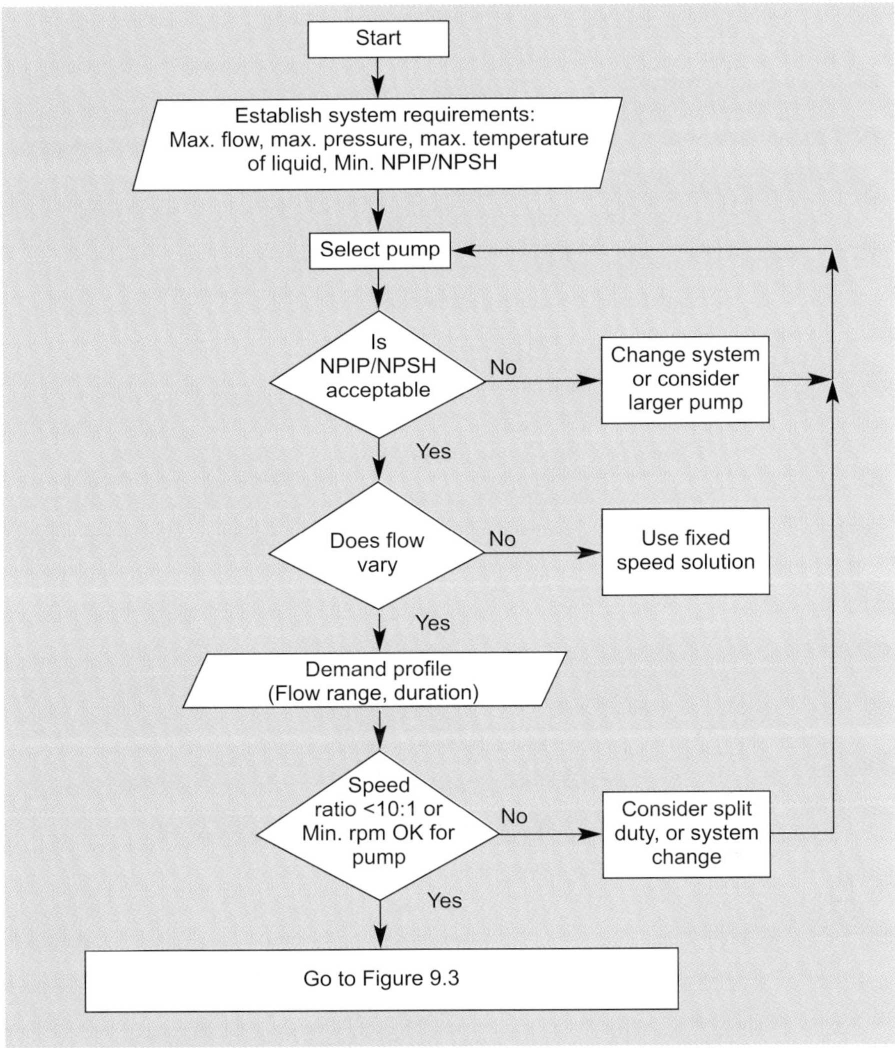

Figure 9.2: Flow chart to assess the suitability of using a VFD in a positive displacement pump system

pumps. These are based on using a variable frequency drive, which will be more efficient than any speed control device, which uses slip to vary the speed. Having established that using a VFD is appropriate, Figure 9.3 is a flow chart for selecting the correct drive and making the life cycle cost calculation. Chapter 12 will help in making the financial justification.

Appendix A2.4 contains information on the factors to be considered in sizing a drive.

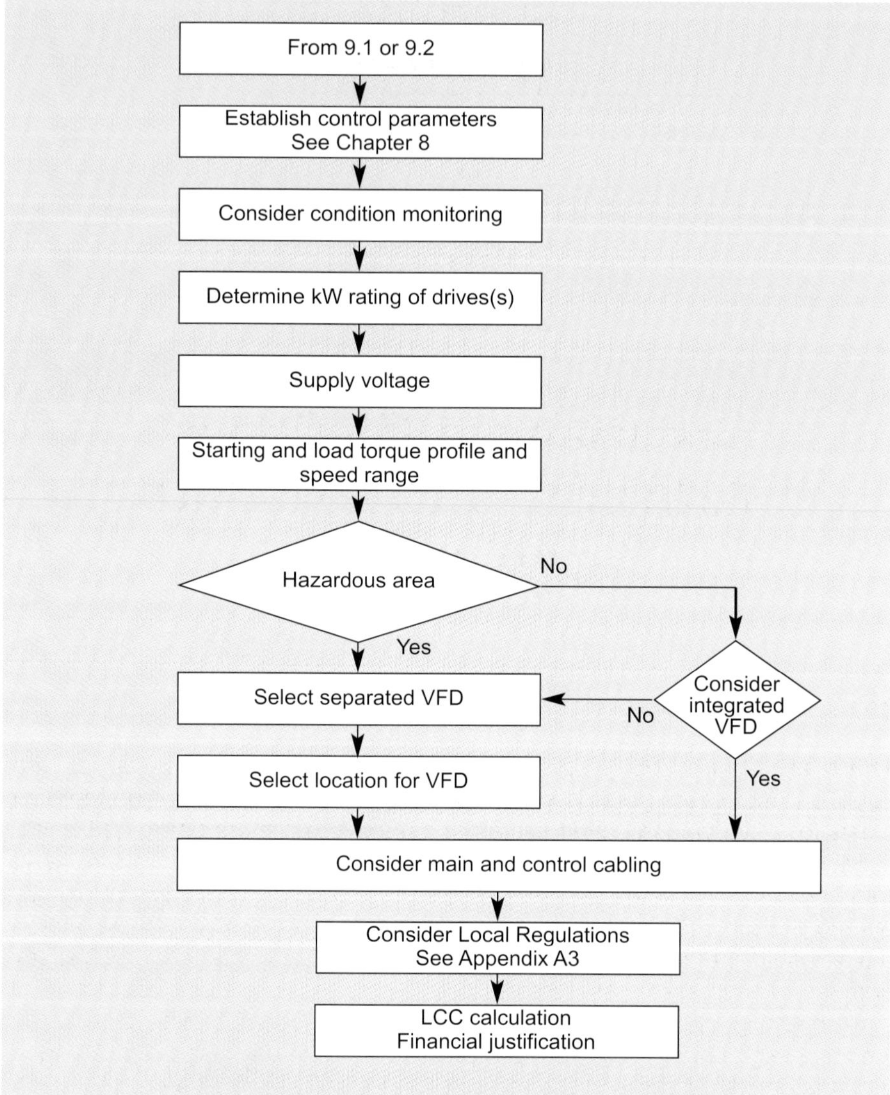

Figure 9.3: Flow chart for selection of the correct drive and financial justification

CHAPTER • TEN

Selection process – retrofitting to existing equipment

10.1 Justification

There are approximately 20 times as many pumps in service as are supplied new every year. It is therefore apparent that a major opportunity exists for modifying installed systems to make them more energy efficient.

Most system designers allow a contingency on the system head required. It is estimated that 75% of pump systems are oversized, usually by more than 20%. It follows therefore that retrofitting variable speed drives could match pumps to actual system requirements more accurately and save considerable amounts of energy.

Also many existing systems use control valves and bypasses, all of which absorb energy not required to satisfy system demands. The fitting of variable speed drives and the removal of control valves and bypasses will save energy and often the payback for the modifications is short. Just consider that a speed reduction of 10% with a rotodynamic pump will save approximately 30% of the electrical energy absorbed.

Pump manufacturers should be contacted to ensure pumps can be run at slower speeds with no detrimental effects, and the motor manufacturers to ensure motors are suitable for use with variable speed drives. There may also be other alternatives; for example if a pump is oversized but operates at a single duty, the impeller diameter can be permanently reduced, which may achieve the same energy reduction as a VSD but at a small cost.

10.2 Motor suitability and de-rating

When considering adding a variable speed drive to an existing motor, care should be taken to match the electrical characteristics of the motor and frequency converter, otherwise the risk of premature failure is introduced into the system.

The following parameters vary from one converter design to another and will affect the drive system performance:

- Total harmonic distortion
- Peak voltage
- Maximum rate of change of voltage
- Switching frequency
- Cable length between the motor and inverter

Early frequency converters produced outputs with a very high harmonic content in the waveform. This resulted in substantial additional heating of motor windings, and therefore motors were de-rated for inverter use. A modern inverter output causes relatively small levels of harmonic current distortion in the motor windings and therefore little de-rating is normally required. Whilst this de-rating will be minimal, it will vary from one motor design to another. It is also dependent on the type of inverter used.

Motor de-rating may be required to compensate for the reduced cooling at lower speeds when the motor shaft mounted fan is not generating the airflow achieved at normal synchronous speed. Motor manufacturers' expertise must be used to determine how much de-rating is required and their recommendations applied. There are no definitive rules that can be applied to all motors.

In general, little or no de-rating is required with most rotodynamic pump drives, however PD pumps with a constant torque characteristic are more likely to require de-rating of the motor for reduced speed operation, although this is usually a lot less than the starting torque.

Typically, motors with a separately driven, fixed-speed, cooling fan, can deliver a constant torque continuously over a wider speed range. Above normal base speed, the drive provides a constant voltage, resulting in a weakening field. In this area the motor torque capability decreases and the motor can at best deliver a constant power. At frequencies well above base speed, the design characteristics of the motor may introduce further restrictions and specific information for the motor should be obtained from the manufacturer.

Figure 10.2 shows the effect on torque capabilities of varying the frequency (speed) of a typical motor, with a power rating based on Class B (80 Kelvin temperature rise).

10.3 Flow charts

To assist with the decision on whether or not to retrofit on existing equipment, please refer to flow chart Figure 10.1.

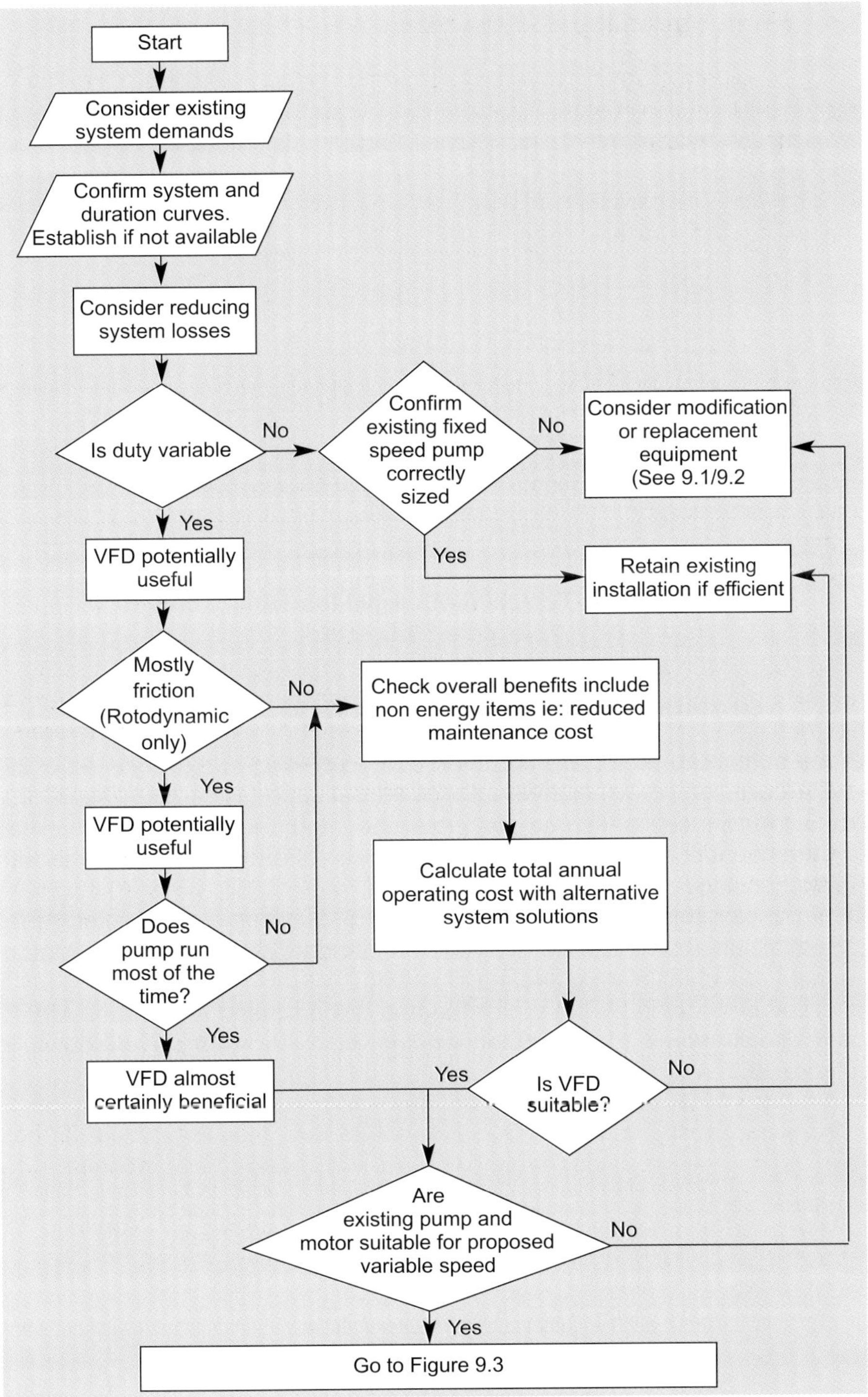

Figure 10.1: Flowchart to assess the suitability of retrofitting a VSD to an existing pump system

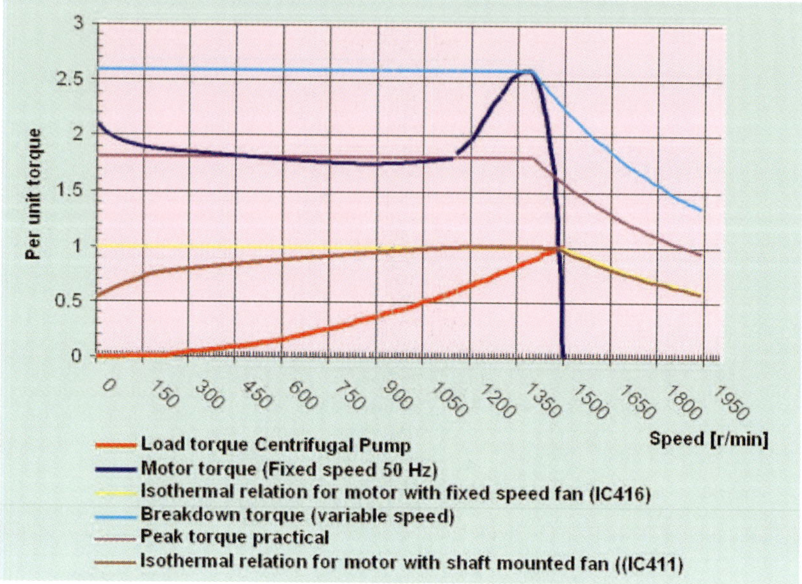

Figure 10.2: Typical isothermal loadability curves four pole 50Hz motor

10.4 Retrofitting a motor mounted VFD

If the pump and motor are not in a hazardous area and the saving of life cycle costs and/or other hydraulic benefits supports the installation of a VFD, in some cases the use of a motor-mounted frequency converter may be the cheapest and easiest way to equip an already installed pump with this technology.

For motor sizes from 1.1 up to 22 kW (1.5–30 hp) motor-mounted VFDs are available which can be mounted on standard three phase AC-Motors without any machining on the motor casing.

They are cooled by the motor fan and are enclosed to IP54/ NEMA 12 and directly wired to the motor terminal box. An example is shown in Figure 10.3.

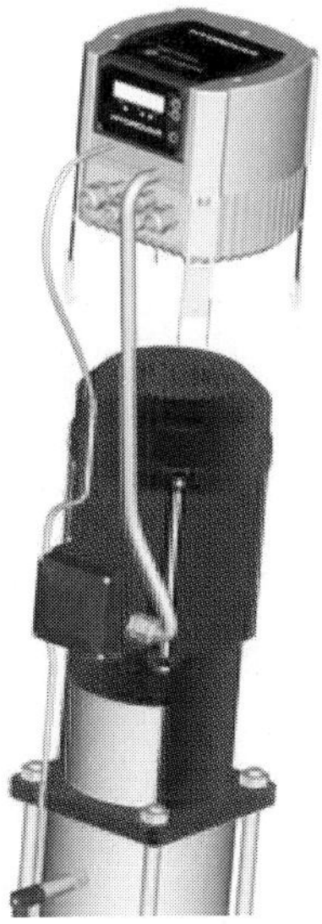

Figure 10.3: Example of retrofitted motor mounted VFD

CHAPTER • ELEVEN

Benefits, drawbacks and operational issues

When designing a system to utilise variable speed technology there will be some benefits that are relatively easy to quantify, some which are less tangible and some potential problems, which must be avoided.

This Chapter provides extensive practical considerations, which will assist in achieving successful VSD installations.

11.1 Tangible benefits to the user

11.1.1 Energy savings

With rotodynamic pump installations, savings of between 30% and 50% have been achieved in many cases by installing VSDs. By adopting all the possible best practices in system improvements identified in this guide, savings as high as 90% have been achieved.

Where positive displacement pumps are used, energy consumption tends to be directly proportional to the volume pumped and savings are readily quantified.

11.1.2 Improved process control

By matching pump output flow or pressure directly to the process requirements, small variations can be corrected more rapidly by variable speed drive than by other control forms, to improve process performance.

There is less likelihood of flow or pressure surges when the control device provides rates of change which are virtually infinitely variable.

11.1.3 Improved system reliability

Any reduction in speed achieved by using a VSD has major benefits in reducing pump wear, particularly in bearings and seals, and by using reliability indices, the additional time between maintenance or breakdowns can be accurately computed.

It is generally considered that wear in bearings and on rubbing surfaces reduces as the seventh power of speed, hence major benefits can be realised by running a pump more slowly.

It therefore follows that the fitting of a VSD has a significant benefit in improving system reliability.

11.2 Additional benefits

The following additional benefits apply specifically to VFDs:

- Modern power electronics are very reliable: typical MTBF figures achieved by manufacturers now exceed 20 years, subject to routine maintenance.
- Easy to retrofit.
- Good supply side displacement power factor at all loads, power factor compensation unlikely to be required, unlike network connected fixed speed motor.
- May possibly be by-passed in the event of failure.
- Open-loop designs can run without any speed sensor. Few pump applications require the degrees of control accuracy where closed loop control is justified.
- No inrush current from the supply network on starting. Soft starting and stopping reduces stress on the motor, pump, coupling and supply network.
- Integrated diagnostics and protection helps reduce and identify faults. VFDs can incorporate several types of dedicated pump and motor protection, (e.g. overload, underload, loss of phase, under/ overvoltage, etc.). With additional sensors the pump can be protected against dry run, cavitation, low flow, seizure etc.
- Many VFDs incorporate control logic: proportional, integral and derivative control loops (three term controller).
- All but the most basic VFDs have serial communications interfaces for networking or for use with over-riding control systems.

11.3 Potential drawbacks of PWM VFDs

- PWM drives can be fitted to most existing motors in Europe and other areas, which use a 400 V network. However this is generally not the case in the US, and other areas, where network voltages exceed 440 V and hence motors with reinforced “inverter duty” insulation are often needed.

- Because many standard drives use a simple diode input bridge, there may be some undesirable harmonic distortion. Attenuation methods are discussed in Appendix A3.3.
- The high rate of switching in the PWM waveform can occasionally lead to problems:
  - The rate of rise of the wave-front can cause electromagnetic disturbances. Adequate electrical screening (screened output cables) and rate of rise (dU/dt) or sinusoidal filters in the inverter output can eliminate this problem.
  - Older motor insulation systems may deteriorate more rapidly due to the rapid rate of change again a sinusoidal filter will eliminate this problem.
  - Long cable runs can cause “transmission line” effects, and cause raised voltages at the motor terminals.
- Voltages can be induced in the shafts of larger motors potentially leading to circulating currents, which can destroy bearings; corrective measures can include:
  - Insulated non-drive end bearings are recommended on all motors over 100 kW (150 hp) output rating.
  - Common mode filters may additionally be required for higher powers and voltages.
- The converter will have losses, the ventilation requirements for the electronics can be an important issue:
  - The life expectancy of the converter is generally directly related to the temperature of the internal components, especially capacitors.
- The converter may require installation in a less onerous environment than the motor control gear it replaces:
  - Electronics are less able to cope with corrosive and damp locations.
  - For a motor in a hazardous area the VFD must be installed in a safe environment, which can lead to long cable runs.

11.4 Operational issues

11.4.1 General precautions

As described earlier in this guide, with rotodynamic pumps, relatively small speed changes can have a major effect on parameters such as power absorbed, NPSH required etc.

Generally VSDs will be used to reduce speed from the nominal. However if there is adequate motor power available the drive may be used to increase speed. In this case there are a number of precautions, which must be considered:

- There will be a change in the noise output from both the pump and the motor as the speed is changed. With higher speed, greater noise and potentially greater vibration will occur.

- If the pump is handling liquid containing abrasive particles, an increase in flow will result in a corresponding increase in abrasion and wear.
- With increasing speed, the user must ensure that the NPSH available at the pump is still at an adequate level to prevent cavitation. This must be checked, since the pressure drop along the suction pipe will increase as the flow/velocity increases, which will cause a reduction in the NPSH available to the pump. The situation is further exacerbated by the fact that the pump will also require more NPSH when running at higher speeds.
- Various mechanical constraints must also be checked. If the speed (and so the power) is increased then bearings, flexible coupling, magnetic coupling, motor, etc must be checked to make sure they can operate with the increased loading.
- Since a speed increase will also affect mechanical seals, the manufacturer should verify that they are still being operated within the allowable region.
- In some instances where shear sensitive liquids/chemicals are involved, it is necessary to limit maximum pump speed.

Lower speed generally benefits energy savings and maintenance costs but the following precautions must be considered:

- In the case of relatively high static head, care has to be taken that the pump is not operated in inappropriate regions. As shown in Figure 4.6, a relatively minor decrease in speed can move the operating point close to shut off head. If the pump is operated in such a region for a long period of time it will have severe consequences for the life of the equipment.
- If pump speed is reduced in some applications, particulates may come out of suspension and cause problems.
- A minimum speed limit applies if pumps have a balance disc or drum (generally multi-stage pumps) as unacceptable wear can occur.
- Self-priming and regenerative pumps will often not prime at low speeds.
- Gas seals have minimum peripheral speed and pump flow rate requirements, check with manufacturer.

11.4.2 Resonance and rotor dynamics

11.4.2.1 Structural resonance

Resonance conditions can result in excessive vibration levels, which in turn are potentially harmful to equipment and environment, particularly with PD, vertical and large rotodynamic pumps.

Pumps, their support structure and piping are subject to a variety of potential structural vibration problems (resonance conditions). Fixed speed applications often miss these potential resonance situations, because the common excitation harmonics due to running speed, vane passing frequency, plunger frequency etc., do not coincide with the structural natural frequencies. For VSD applications, the excitation frequencies become variable and the likelihood of encountering a resonance condition within the continuous operating speed range is greatly increased.

Pressure pulsations are the common excitation mechanism. These pressure pulsations may be further amplified by acoustic resonance within the pump or the adjacent piping.

11.4.2.1.1 Precautions There are a number of analyses that can be performed in order to predict potential resonance situations:

- Simple hydraulic resonance calculations.
- Passing frequency analysis.
- Using finite element analysis to investigate structural resonance.
- Modal testing of the physical machine can supplement the regular vibration test. Very often a pump intended for variable speed operation will only be tested at one single speed.

11.4.2.2 Rotor dynamics

The risk of the rotating element in the pump or motor encountering a lateral critical speed increases with the application of a VSD. Lateral critical speeds occur when running speed excitation coincides with one of the rotor's lateral natural frequencies. The resulting rotor vibration may be acceptable or excessive, depending on the modal damping associated with the corresponding mode.

Additionally, drive-induced torque harmonics may result in resonance conditions with torsional rotor dynamic modes.

11.4.2.2.1 Precautions

- Lateral and torsional rotor dynamic analyses allow the calculation of rotor natural frequencies under operating conditions over the entire applicable speed range. Lateral modes are typically damped and the acceptability of a particular design depends on a combination of available modal damping and separation between natural and excitation frequencies.
- Torsional modes have little damping. The acceptability of the design may be proven by means of a forced response analysis and comparison of calculated stresses to allowable stresses.
- Three basic corrective actions, or a combination thereof, can be applied if a resonance situation exists and associated vibration levels are excessive:

- Locking out of certain speeds or speed ranges from the continuous operating speed range can be achieved in the drive software. However this may limit the operational flexibility.
- Detuning a resonance condition by means of changing the structural natural frequency or changing the excitation frequency.
- Reducing excitation levels by means of improving balance, alignment etc. or by means of changing impeller/diffuser or impeller/volute configurations.

11.4.2.3 Resonant vibrations in vertical pumps

Variable speed vertical pumps are more likely than horizontal machines to exhibit operational zones of excessive vibration, because their lower natural frequencies are more likely to coincide with running speeds. Some specific guidance is therefore provided in this Chapter.

Small, vertical close-coupled and multistage pumps do not normally show this type of problem.

Most manufacturers of large vertical pumps and motors have in-house technology to determine the natural reed frequency (NRF) and will do this job by job on engineered vertical pump sets.

It is normal practice to design the driver support (discharge head) of vertical pumps, such that the NRF of the pump/motor unit has a separation margin of 20% above the maximum operating r/min, or 20% below the minimum operating r/min. The motor NRF is needed to make this calculation. If this criterion is not satisfied it may be possible to select a pump with a different synchronous full speed; as an example changing from six pole to four pole or six pole to eight pole, to achieve the desired non-resonance band of operation. The specifying engineer should consult with the pump manufacturer prior to specifying motor synchronous speed and the pump speed ranges, when the vertical pump is driven by a variable speed driver.

Another way of avoiding the pump running at the NRF is to lock out the problem speed zone of operation on the variable frequency drive (VFD) control. The pump may have different NRFs in the x and y directions, so it is necessary to avoid a band of speeds as shown in Figure 11.1. This may limit the pump's useful range of operation: therefore, it is usually preferable if possible to change the pump/motor structural stiffness to move the resonance band away from the desired operating zone.

The influence of the foundation stiffness on the NRF must be considered. It is desirable to have a foundation that will create less than 5% of the total deflection of the pump/driver structural elements. Typically a reinforced concrete foundation will achieve this.

If the NRF is close to the maximum range of operating speed (Figure 11.2), the unit usually may be stiffened. However, there is a limit to the pump manufacturer's ability to stiffen the unit's structure. Procuring a stiffer vertical motor may not be feasible. However, by providing a stiffer motor support (fabricated discharge head) it may be possible to raise the pump's NRF above maximum operating speed.

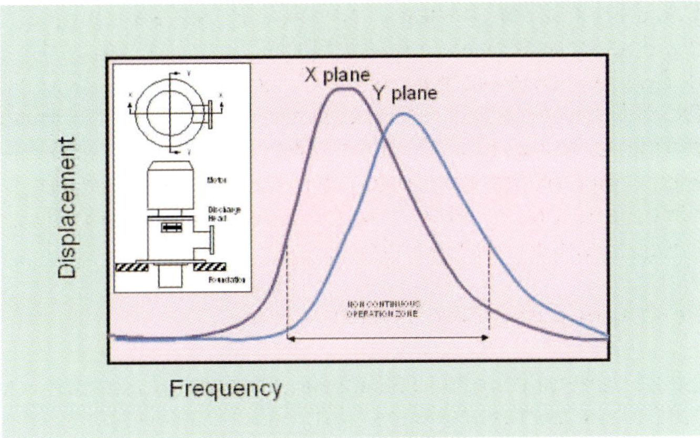

Figure 11.1: Typical vibration signature of discharge head/motor structure

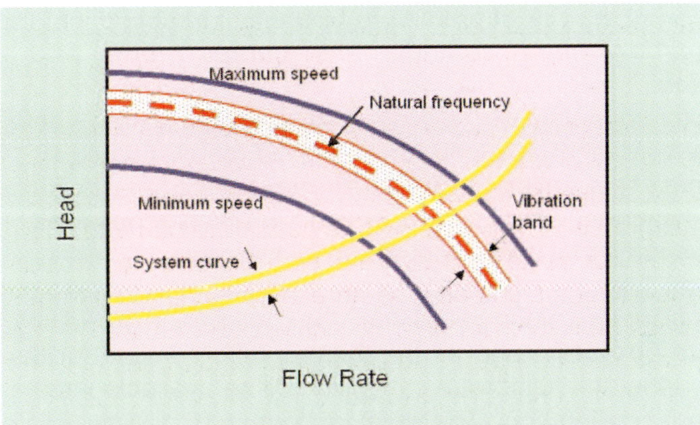

Figure 11.2: NRF at a frequency close to maximum operating speed

If the NRF is close to the minimum range of operating speed (Figure 11.3), the unit may be made more flexible. This will lower the natural frequency below the operating range. During start-up and shutdown, the unit will pass through its natural frequency and vibration may occur. However, this vibration should not be damaging due to its very short duration of time.

11.5 Power drive system integration

A motor with its control becomes a power drive system, which may be fixed or variable speed design. This section considers the integration of a variable speed system.

When considering using a variable speed drive with a motor, great care must be taken to match the electrical characteristics of the motor and

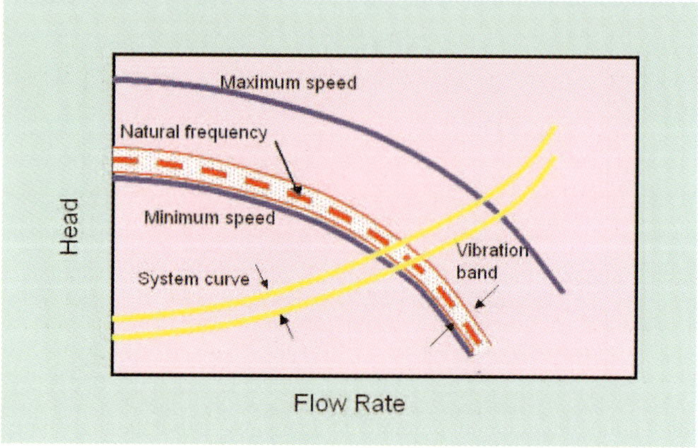

Figure 11.3: NRF at a frequency close to minimum operating speed

drive, otherwise the risk of premature failure is introduced into the system. ***The evaluation must consider the pump, motor and drive in combination.***

The motor manufacturer will generally be able to give details of the acceptable operating envelope for a specific motor, Figure 10.1 shows a typical curve. This envelope may well vary from frame to frame, and is especially important when a motor is used in a hazardous area. There are no definitive rules that can be applied to all motors.

In general, little or no de-rating is required with rotodynamic pumps.

It must also be remembered that the isothermal guidelines apply to ***continuous operation*** and as the thermal time constant of the motor is reasonably long, the full torque capability will be available intermittently for starting and acceleration. The level of torque is directly related to the current passing, and is therefore governed by the converter.

With a high inertia load, such as a pump with a flywheel, a VFD is able to control the starting current and allow the accelerating torque to be sustained for longer than for a direct-on-line start.

PD pumps with their constant torque characteristic are more likely to require special consideration for reduced speed operation. In addition, the levels of starting torque required need to be considered, especially with progressive cavity pumps.

11.5.1 Operating motors above base speed

In some instances operating a motor beyond base frequency is possible if the consequences are carefully examined as shown in the following example.

If a PD pump requires 200 Nm torque to breakaway, and 100 Nm torque in the operating range from 500 r/min to 3000 r/min pump speed, i.e. around 31 kW absorbed power at maximum speed, it is possible to select

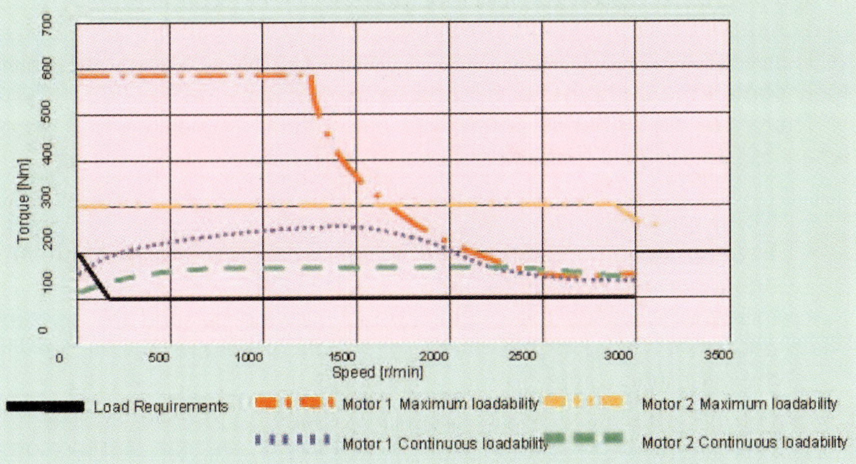

	Motor 1	Motor 2
Rated Power & Speed	45 kW (60 hp) at 2960 r/min	30 kW (40 hp) at 1470 r/min
Current at breakaway (200 Nm)	120.4 A	66.6 A
Running current at duty (100 Nm)	61.5 A	55 A
Inverter rating	70 kVA	40 kVA
Relative cost Motor + Drive	144 %	100 %

Figure 11.4: Motor load capability curves

a two pole motor, which will run in the constant flux area. An illustration of the above is shown in Figure 11.4.

A typical 37 kW 400 V two pole 50 Hz self ventilated motor will run outside of its thermal capabilities below 1500 r/min, so a 45 kW motor would normally be selected. (This is substantially less costly than fitting a fixed speed fan to the 37 kW option.)

The 45 kW motor draws around 62 Amps when delivering 100 Nm at 3000 r/min, but will require around 120 A to develop 200 Nm at breakaway.

An alternative selection is to use a 37 kW four pole motor with a normal synchronous speed of 1500 r/min, at which it is capable of running at 37 kW at 50 Hz 400 V delivering 240 Nm and drawing 70 A.

This motor will also operate at 400 V, 100 Hz, (3000 r/min), delivering 100 Nm, while drawing around 55 A. A current of around 65 A is required to develop 200 Nm at breakaway.

This means that the converter current required for the two pole solution is double that required for the four pole solution. A substantially smaller and cheaper converter can therefore be used with the four pole motor.

This enables a lower rated drive to be used, while still ensuring satisfactorily control the pump/motor over its desired duty range.

There are a number of points that also need to be taken into account when considering such solutions:

- Noise will be increased with speed.
- Bearing life or greasing intervals may be reduced, or improved fit bearings may be required.
- In extreme cases the higher speed may be beyond the maximum rotational speed of the motor, or a critical speed may occur within the operating range.
- Another important point that also needs to be considered when using a motor with lower base speed to deliver high starting torque, is whether the torque required at full speed can de provided by the motor, because not only does its rated torque fall inversely with frequency but the break down torque falls inversely with the square of frequency, either of which can be the restricting parameter.

11.6 Low base speed motors

For applications requiring operation with relatively slow base speed motors (10 or more poles), the limited number of slot combinations between the stator and rotor can lead to an acoustic resonant condition, and the associated very high levels of narrow band noise.

Many motor manufacturers will offer special windings using four pole or six pole motors with a reduced nominal frequency to avoid this phenomenon.

For a European 400 V electricity supply network, a requirement to deliver full output at, for example 600 r/min, could be achieved with a 10 pole motor design, wound for 400 V at 50 Hz, or with a six pole motor wound for 400 V at 30 Hz.

Equally, because many motors are wound 400 V Delta/690 V Star (Wye), it is possible to provide full flux to a six pole design by connecting the motor in star and feeding with 400 V at 29 Hz base.

11.7 Motor design

Motors can be designed for different features of starting torque, starting current and slip.

Standard motors designed to IEC standards are designated NY, N, H and HY; while motors designed to NEMA carry letters A, B, C and D. Note that the specifications are not totally compatible, but are combined in Table 11.1 for convenience.

Table 11.1: Motor designs

DESIGN to IEC	NY	N	H	HY
DESIGN to NEMA	A	B	C	D
FEATURE				
Starting Torque	Normal	Normal	High	High
Peak Torque $T_{(max)}$	High	Medium	Medium	Low
Starting Current	High	High	Low	Medium
Slip at full load	Low	Medium	Medium	High
Used for	Fast starting	Standard	Heavy loads or weak supply networks	High inertia and reversing drives

For voltage source PWM inverter operation, design NY or A is the best suited, while design N or B will normally give satisfactory service. For other motors consult the manufacturer.

Specially designed inverter fed motors are offered for a number of specific low power requirements, such as high speed operation, where the base speed may be increased from 3000 r/min to 4000 or 5000 r/min. However, special designs of inverter fed motors are more widely used for higher powers and will generally carry a special name and rating plate designating the permitted duty range.

CHAPTER • TWELVE

Financial justification

12.1 Life cycle cost (LCC)

The initial cost of pumping equipment is often a very small part of the total life cycle cost. An LCC analysis is therefore a very appropriate way of comparing different technical alternatives in the design of a pumping system and making a financial justification.

A very well documented LCC guide has been published and is available from the Hydraulic Institute and Europump; www.pumps.org and www.europump.org (Ref ISBN 1-880952-58-0). This guide explains how the operating costs of a pumping system are influenced by system design and shows in detail how to use a life cycle cost analysis to make comparative cost assessments. Many case studies have been developed in the guide to train the engineer in the LCC method.

The life cycle cost represents the total expenses to purchase, install, operate, maintain and repair a pumping system during its lifetime. Down time and environmental costs are also considered.

The LCC equation and its components are shown below.

$$LCC = \text{The sum of } (C_{ic} + C_{in} + C_e + C_o + C_m + C_s + C_d + C_{env})$$

Where:

LCC = Life Cycle Cost
C_{ic} = Purchasing cost (pumps, systems, pipes, auxiliaries)
C_{in} = Installation and commissioning cost
C_e = Energy cost over the expected process or equipment lifetime (including pump, operation of the system, control/command and all auxiliaries services)
C_o = Operating cost (labour for normal system operation)
C_m = Maintenance cost (for scheduled / unscheduled maintenance operation)

C_s = Cost of lost production
C_d = Cost of final decommissioning and environmental site reconstitution
C_{env} = Environmental cost (pollution by pumped liquid or auxiliary equipment)

The way in which a VSD influences the different components of LCC is summarised in Table 12.1.

Table 12.1: Influence of VSD on LCC

Type of Cost	Cost Influence of using a VSD	Explanation
C_{ic}	Can show cost saving	Cost savings on valves and bypass lines can more than off-set the cost of the VSD. See capital cost saving in Chapter 12.2.
C_{in}	Can show cost saving	The VSD system may need a little more time for VSD cabinet installation and cabling, however this is off-set by elimination of control valve and starter.
C_e	Decrease	Correctly applied, large savings can be achieved on energy by operating the pumps at reduced speed to match system demand. A VSD will also often allow the pump to operate closer to its best efficiency point.
C_o	No change	Operation cost is the same. Modern PLCs are well designed for VSD operation.
C_m	Decrease	Maintenance costs are greatly reduced when a VSD is used instead of a valve. A VSD will also reduce the loads associated with Direct-on-Line or Star/Delta (Wye/Delta) starting. Very little maintenance is necessary for the VSD system itself.
C_s	Decrease	Down time cost is expected to be reduced, due to the fact that a VSD system can be started and stopped slowly and that the system operates at reduced speed during long periods.
C_d	No change	Equivalent cost to fixed speed pump sets.
C_{env}	No change	Equivalent cost to fixed speed pump sets.

12.2 Capital cost savings

When designing and installing a new pump system, the capital cost of the VSD can often be off-set by eliminating control valves, by-pass lines and conventional starters, as explained in this Chapter.

12.2.1 Elimination of control valves

Control valves are used to adjust pump output to suit varying system requirements. Usually a constant speed pump is pumping against a control valve, which is partially closed for most of the time. Even at maximum flow conditions, a control valve is normally designed to be 10% shut, for control purposes, and hence a considerable frictional resistance is applied. Energy is therefore wasted overcoming the added frictional loss through the valve.

By installing a variable speed drive, the output of the pump can be varied to match the system requirements without throttling the pump. The losses associated with a throttling valve can therefore be eliminated together with the valve itself. In many cases, where the system losses are difficult to calculate, safety margins will lead to over-sizing of pumps. By using variable speed drives in a system, such over-sizing can be compensated for, by reducing the speed of the pump instead of throttling the flow.

The cost of the mechanical installation of the valve, the associated pipework requirements and the electrical wiring, will be replaced by the cost of purchasing and installing the VSD. It is generally accepted that the installation costs of a control valve can be at least twice the purchase cost of the valve; hence considerable sums can be saved if this item is not required.

If the liquid handled requires special materials, the cost of the control valve can be even more substantial.

12.2.2 Elimination of by-pass lines

All fixed speed centrifugal pumps have a minimum flow requirement. If the pump is operated at flow rates below the minimum for extended periods of time, various mechanical problems can occur. If the flow requirements in a system can drop below this minimum flow capacity, it is necessary to install a constant or switched bypass in order to protect the pump.

With a constantly open bypass line, excess energy is absorbed in continuously pumping the minimum flow down the bypass even though bypass flow is only needed in cases of low system flow requirements.

In a switched bypass, the bypass line opens, usually by use of a solenoid valve, when the system requirement is below minimum flow conditions. In this case, excess energy is used pumping the liquid down the bypass only when the bypass is in operation.

If a pump is controlled by a variable speed drive the need for the bypass may be eliminated by using an intelligent control system, or at least the bypass flow may be greatly reduced.

Capital cost savings are therefore possible if the bypass line is not required. In addition to the piping, savings will also be made on switched bypass systems, by eliminating the need for solenoid valves, with associated pressure or flow monitoring devices and wiring.

12.2.3 Elimination of starters

Fixed speed pumps always require an electrical starter. However, most variable speed drive units work without a starter and will also provide a soft start.

Therefore the capital cost of a starter and the associated wiring can also often be saved. Again it is generally accepted that the installation cost of a starter is at least twice the capital cost of the item. Hence the capital and installation cost saving of the starter can be offset against the cost of the VFD.

CHAPTER THIRTEEN

Case studies

Introduction

A number of VSD case studies are presented in this Chapter to illustrate the potential savings possible, and also the less quantifiable improvements in pump and system reliability. For example, in one case the greatest benefit of using a VSD was not the cost saving, but that the process (irrigating a golf course) was carried out in a more effective and controlled way.

The industries in which the applications were based are described along with the ways in which the benefits were achieved. This should allow similar opportunities to be found, whatever the process or type of industry.

The amount of the saving will be dependent on the cost of energy for the user. Where the study recorded energy cost per kilowatt hours (kWh), it is stated.

The value of the Euro and US dollar are considered sufficiently close, for the purposes of the case studies, to make conversion unnecessary, the monetary values are expressed as €/$.

Where gallons are used in case studies, they are US gallons.

Actual company references for equipment suppliers or user industries are intentionally excluded.

Case study 1: Variable speed drives on water pumps supplying a ring main at a domestic appliance factory

(Based on *Energy Efficiency Best Practice Programme*, case study 88)

Case study summary

Energy savings were achieved by using variable speed drives (VSDs) to control two water pumps supplying the ring main in a domestic electric appliance factory. Previously control was by pressure-operated bypass valve, with pumps oversized for normal operation.

Other potential applications

Any applications which require varying flow rates, where present control method is by valves.

Investment cost

€/$ 11 350.

Savings achieved

191 000 kWh/year worth €/$ 9700 (The company electricity price was €/$ 0.05 per kWh).

Water distribution system

The process water distribution system at a domestic electrical equipment manufacturer consisted of a 150 mm (6 inch) cast iron distribution ring main, which was fed from two sources:

1. Mains water from the local water supply company (termed raw water in this study) was pumped into the ring main from a 910 000 litre (240 000 gallon) storage tank using one of two pumps, each driven by a 15 kW (20 hp) induction motor. The pumps were used on a duty/stand-by basis and both were capable of providing much higher pressures and flow rates than required.
2. Water from the company's own borehole was pumped directly into the distribution main by a submersible borehole pump with 30 kW (40 hp) motor.

Original control method

With the raw water pumps, control of both the ring main pressure and the pump flow rates was originally achieved by mechanical means. Pressure in the ring main was controlled by a pressure-operated bypass valve, which diverted excess water back into the storage tank. The system caused the raw water pump to be fully loaded at all times.

When using the borehole pump, the allowable flow rate was governed by the company's water abstraction contract. To maintain flow rates within this limit a partially closed valve between the pump and ring main was used to throttle the flow, with considerable energy wastage.

VSD installation and commissioning

Borehole pump

Installation of the borehole pump VSD allowed the flow rate to be controlled by changing the pump speed, thus removing the need for the throttle valve, and thereby reducing the power consumption of the motor.

The VSD was rated at 27 kW output. This unit was selected as it was envisaged that the motor loadings would not exceed 27 kW when operating under the new regime.

Speed control of the pump was achieved by manual adjustment of a potentiometer which provided the VSD with a signal representing the desired pump speed, and hence giving the required flow rate. Automatic control was considered and then rejected as being unnecessary; changes in pump speed were only required to compensate for pump wear or to meet new abstraction allowances.

Raw water pump

The VSD installed was rated at 15 kW, which was also the power rating of each of the raw water pumps.

Control of the ring main water pressure was achieved by varying the speed of the pumps in response to a signal from a pressure transducer, which was fitted in the ring main. This pressure transducer was fitted along with the VSD as a complete package to enable automatic control to be achieved.

Energy saving

Monitoring was initially performed with the system operating under the original conditions and then again under VSD control.

Recordings of energy consumption from the monitoring programme revealed that substantial energy savings had been achieved by each of the installations.

The borehole pump showed a reduction in unit consumption of 79 300 kWh/year, which was equivalent to a 30% saving.

The raw water pump produced even larger savings of 111 750 kWh/year corresponding to an 88% reduction in consumption.

A combined saving from the two installations of 191 050 kWh/year was therefore achieved. This was equivalent to an overall reduction in consumption of 48%.

Economic analysis

Savings achieved by the installation of the two VSDs have been substantial, resulting in short payback periods. Installation of the borehole VSD cost €/$ 6750. The energy saving achieved was worth €/$ 4000 per year, resulting in a simple payback for this installation of 19 months.

The raw water pump VSD cost was €/$ 4600. Comparing this capital cost against savings of €/$ 5700 achieved by the installation gives a simple payback of 10 months.

Combined project savings were €/$ 9700 per year, giving an overall payback period of 14 months.

Case Study 2: VSD in a wastewater pumping scheme

Case study summary

Energy cost and noise were reduced by replacing fixed speed equipment with a VSD so that pump rate was at the minimum necessary therefore reducing pipe friction losses. Whilst actual monetary figures are not available it is believed the payback period was less than two years.

Potential users

Most waste water pumping schemes.

Savings achieved

€/$ 131 000 over a 20 year period.

Existing installation

The station was of the traditional separate wet/dry well design. Wastewater was pumped to the treatment works some 2.5 km (1.5 miles) away. Control was by simple on/off switching of the motors, with the level monitored with a mechanical float.

Installation of VSDs

To monitor the wet well, ultrasonic level instrumentation was installed. This generated a speed reference signal, which increased or decreased the pump speed and also allowed the start and stop levels to be selected. This resulted in the motors running between 73% and 88% duty, which was the optimum speed range for energy efficient operation of the pumps.

Energy efficient motors were also introduced to the system to further improve energy efficiency, as the benefit of high efficiency motors over standard becomes even greater in variable speed operation.

Economic analysis

The old pumps were estimated to transfer 14 m^3 of liquid per kWh. After fitting new pumps and drives, but still running at the same speed as the old pumps, this increased to 21 m^3 per kWh. With the drives interfaced to the analogue reference signal, this was further improved to 30 m^3 per kWh. Monitoring of this value can indicate pump wear or blockage, enabling the service intervals to be optimized.

The efficiency of the wastewater pumping station has more than doubled since the new pumps and VSDs were installed. It is predicted that the savings will be at least €/$ 131 000 in electricity costs over a 20-year life span.

The total electricity consumption per year fell by 53% with 33% coming from the change of pumps and motors and 20% from the introduction of the VSDs.

Other benefits

The installation has reduced noise, which is of importance in the residential area where the station is located.

Case study 3: Variable speed drives on golf course irrigation pumps

Case study summary

More effective irrigation was achieved while saving water, electricity, maintenance and man hours over the manual and fixed speed systems.

Other potential applications

Any applications which requires fixed pressure or fixed flow rate, even though the pipe system changes and varies the friction or static head losses.

Problems with fixed speed irrigation systems

Output requirements vary when different parts of the sprinkler system are in use. Control valves are often used to maintain pressure or regulate flow, with inevitable wastage of energy.

There can be excessive usage of water and uneven application.

Pressure surges, caused by on-off control with long pipe runs, lead to burst pipes.

Benefit of using variable speed pumping

Three different golf courses all reported a successful outcome with VSD pumps, but emphasized different benefits.

A golf club in the London area retrofitted 7.5 kW VFDs to existing fixed speed pumps, each with pressure transmitter and flow meter. The pressure transmitter sends information on water pressure to the drive, which adjusts the speed of the motors to maintain the correct water flow to keep the grass healthy and avoid waterlogging. This keeps the use of both water and power to minimum and the flow meter helps the club to keep track of its water. Maintenance costs are also kept low as wear and tear on pumps is reduced.

The system controls 450 pop-up sprinklers in three zones, each zone with its own pump. The control panel can be set to operate sprinklers in several combinations and for varying times, ensuring that the turf get the correct amount of water.

The automatic watering has not only improved the condition of the fairways but also saves two days work over the manual method. The system also allows fertilizer to be watered-in in spring and keeps the ground in the best condition for spiking equipment to work most effectively without pulling out lumps of turf and being damaged itself on hard unwatered ground.

A Southern Californian golf course replaced fixed speed equipment, comprising three 55 kW (75 hp) pumps and one 30 kW (40 hp) jockey pump, with three 45 kW (60 hp) and one 4 kW (5 hp) jockey pump.

This avoided up to four line breaks a day by eliminating the pressure surges.

Annual electricity cost was reduced by 46% (€/$ 25 000).

A rebate from the power company of $ 100 for every VFD hp was granted.

Computer control with flow and pressure measurement allows precise pressure control for specific irrigation tasks and better management decisions. This is important with the area's water shortage.

A golf course in the Palm Springs area of California replaced three 90 kW (125 hp) pumps, one 55 kW (75 hp) pump and one 22 kW (30 hp) jockey pump, with three 55 kW (75 hp) pumps, one 30 kW (40 hp) pump and one 5.5 kW (7.5 hp) jockey pump, thus almost halving the pumping power. All replacement pumps were variable speed.

Typically there is a 100 000 kWh saving per month, which with an energy cost of €/$ 0.09/kWh corresponds to a €/$ 9000 cost saving. This represents between 30 and 50%.

A rebate of €/$ 6000 was obtained from the power company for use of VFDs.

Pump and motor life were improved and control valve maintenance eliminated.

Case study 4: Variable speed drives on heating, ventilating, air conditioning (HVAC) system pumps

Case study summary

Large energy savings were achieved by fitting VFDs on pumps previously allowed to run uncontrolled at their maximum flow, saving pumping cost and heating energy.

Other potential applications

Heating and air conditioning systems which see varying heating load, i.e. almost all of them.

Savings achieved

€/$ 110 000 per annum (including gas & electricity)

Installation and operation details

Variable frequency drives have been fitted on the pumps for chilled, hot and condenser water and also medium pressure water for the boiler room. The installation was at a large, international private bank in London. One 4 kW, one 22 kW and two 75 kW drives were retrofitted.

Sensors in the pipework, connected to the drives, regulate the speed of the pumps to ensure optimum usage of the water's energy content. The system is coordinated with the building management system, enabling easy control of the indoor climate.

The pumps are now running at the speed needed by the process. This is generally about 40% of the maximum speed, at which the pumps previously ran continuously. Before fitting the VFDs, water would be pumped out at 40°C and return with no temperature drop. Now the return is 32°C. The overall energy saving in gas and electricity in this case is reported to be 90%. The lower pump speed is also reducing pump maintenance cost.

Case study 5: Upgrade of an oil refinery's pumping systems including installation of variable speed drives

Case study summary

At a San Francisco refinery, installing a VSD on a product transfer pump saved €/$ 120 000 per annum and on a primary feed pump saved €/$ 220 000. Vibration was reduced and mechanical seal and bearing failures have been eliminated. There was no investment cost to the refinery but savings were shared with the contractor.

Other potential applications

Any applications where the pump is sized for an intermittent maximum flow rate but runs mostly at a reduced (but variable) rate.

Investment cost

The energy services contractor agreed to install the VSDs and upgrade the equipment at no charge to the refinery, but took a share of the savings. Their total investment was €/$ 1 200 000.

Savings achieved

€/$ 340 000 per annum from the VSDs and €/$ 750 000 from the total project.

Payback period

Overall about 18 months, but this was not applicable to the refinery who gained immediately with their share of the savings.

Installation and operation details

Conversion of the refinery's vacuum gas oil plant to a diesel hydro treater (DHT), left the pumps grossly oversized. Several pumps were operating frequently at 40% of best efficiency point, resulting in low hydraulic efficiency, excessive vibration and seal or bearing failure about once a year.

The full range of upgrades consisted of:

- Installing VSDs on the 1650 kW (2250 hp) primary feed pump and the 500 kW (700 hp) product transfer pump.
- Replacing the internal elements on the 1650 kW (2250 hp) secondary feed pump and on a 300 kW (400 hp) power recovery Turbine (PRT).

- Changing operating procedures for the main 3700 kW (5000 hp) and 3000 kW (4000 hp) back-up pumps.

The installation of the VSDs on the primary feed pump and product transfer pump gave energy savings by reducing losses through flow control valves.

The energy saving from the use of VSDs was 500 000 kWh per month. The resizing of the PRT and secondary feed pump, along with a more energy efficient operating procedure for the back-up pumps, saved another 500 000 kWh per month. Cost savings shared by the refinery and contractor were €/$ 340 000 per annum from the variable speed pumps and €/$ 750 000 overall.

The demand charge previously levied on the DHT process was eliminated.

Since the upgrades there have been no seal or bearing failures and process control has improved.

It should be noted that a VSD was not considered appropriate for all the oversized pumps. If the duty is not varying then resizing the pump, e.g. new impeller and diffuser element, reduced impeller diameters or even a new pump, will usually give greater life time cost saving and better pay back than a VSD.

Case study 6: Variable speed drive on a circulating pump in a batch process

Case study summary

A reduction in seal failures and maintenance cost savings were achieved by using a variable speed drive (VSD) to control a circulating pump in a batch process for a major chemical manufacturer. Additional benefits were energy savings and increased productivity. This was a retrofit application where a constant speed pump previously controlled the process.

Other potential applications

Pump applications that have high maintenance costs due to frequent seal failures caused by dry running or pump cavitation. Pumps that experience bearing failures due to operation at very low flows for extended periods. Any application that utilizes a throttle valve to control a process.

Investment cost

The total installed cost was approximately €/$ 25 000.

Savings achieved

Total savings were €/$ 62 500 per year.

Payback period

Payback was achieved in less than 5 months.

Installation and operation details

Original control method

This manufacturer produces fragrance-based chemicals that are used in various consumer products such as detergents and soaps. Since some of the chemicals being pumped were highly corrosive, the pump was constructed with expensive duplex stainless steel.

The process utilized an end suction centrifugal pump in a distillation tower for circulation. The pump recirculated process liquid through a heat exchanger and back through a reboiler to keep it hot. Running at a fixed speed of 1170 r/min the pump was emptying a suction tank to fulfill a process need in a batch application. As the level of liquid in a steel pot was distilled off, the suction level would drop from 100% to virtually 0%. As a result the pump was experiencing dramatically varying suction conditions. Under normal operating conditions the pump would function well; however, during the last hour of each process the pump experienced extreme cavitation due to the low suction tank level. With these severe operating conditions the pump was experiencing on average a mechanical seal failure every 17 days.

Intelligent VSD based solution

A VSD was installed which had custom pump protection and control software that resides directly on the VSD microprocessor. The custom pump software, with feedback from instrumentation, provided unattended control of the pumping system and protected the pump from operating under damaging conditions. The VSD automatically reacted to process system changes and protected against problems such as dry running, cavitation, operation below safe minimum flow or against a closed suction or discharge valve.

In this application the VSD reacts to a lowered suction level condition by gradually slowing the pump down to prevent cavitation and damage to the mechanical seal. Since installing the VSD, the mean time between failures (MTBF) of 17 days has increased to almost no failures. Another major benefit is that productivity has increased between 5 and 10%.

Energy savings

Monitoring was initially performed with the system operating under the original conditions and again with the VSD control. The original control method utilised a fixed speed 30 kW (40 hp) 1200 r/min motor. The VSD based solution currently uses between 5 kW (7 hp) and 7.5 kW (10 hp) while operating between 550 and 700 r/min. Based on a €/$ 0.04 per kWh, this results in an annual saving of €/$ 8500.

Economic analysis

Savings due to reliability improvements have been dramatic and resulted in a very short payback period. The virtual elimination of mechanical seal failures produced average annual savings of €/$ 54 000 per year and resulted in the elimination of downtime. The original fixed speed units proved to be oversized. VSD operation resulted in an additional saving of €/$ 8500. Total savings on this retrofit application were €/$ 62 500. The simple payback was under 5 months on an investment of €/$ 25 000.

Additional savings were achieved due to a 5–10% increase in productivity; however, the chemical manufacturer would not make these savings available for publication.

Case study 7: Variable speed drives on a paper stock transfer pump

Case study summary

Installing a VSD on a low-density stock pump saved energy and reduced pipework and valve maintenance cost by eliminating the flow control valve. The total annual saving was €/$ 30 000.

Other potential applications

Applications where flow control by valve can be eliminated (or minimized) by the use of variable speed control.

Investment cost

€/$ 17 500

Savings achieved

€/$ 17 000 energy cost and approximately €/$ 13 000 maintenance and repair costs, i.e. a total of €/$ 30 000 per annum.

Payback period

Less than one year.

Installation and operation details

The paper mill relied on a valve to control flow from the 150 kW (200 hp) fixed speed pump on the thermo-mechanical pulp (TMP) mill's storage tower. Operators throttled a 250 mm (10 inch) control valve in the pump discharge to maintain a set level in the TMP low-density storage tower. Often the valve had to be nearly closed due to process demands. Not only did this waste considerable amounts of energy with the high head loss across the valve, it also resulted in severe cavitation across the valve, leading to valve and pipe damage. The pump also had to operate in a low flow region where pump reliability suffered.

A VSD was installed with sensors to monitor system parameters such that the controller could vary speed to maintain a set level in the storage tower and avoid an undesirable operating condition for the pump. The control valve was removed, being no longer required for flow control.

As a result, motor speed was reduced from 1150 r/min to generally around 450 r/min, saving about 75 kW (100 hp) which was equivalent to €/$ 17 000 per annum. Eliminating the control valve and its associated repairs saved a further €/$ 13 000 per annum.

Case study 8: Benefits given by the use of environmental temperature control using reciprocating PD pumps driven by VSD

Case study summary

Reciprocating positive displacement pumps were used for the application of temperature control in a paper mill, where the requirement was for a variable flow rate of demineralised water.

To cool high ambient temperatures for a number of applications, the technique of atomizing water into fine droplets was employed.

The performance required was from 0.1 l/min to 3.0 l/min at pressures up to 80 bar. The duty involved cooling the environment surrounding the paper forming process to prevent premature drying.

Installed details

The PD pumps chosen were powered by an electric motor controlled by an integrated electronic variable frequency drive.

The water, vaporized at the nozzles, lowers the temperature by absorbing its latent heat of vaporization.

By linking a temperature sensor via a programmable logic controller (PLC) to the variable frequency drive it is possible to control the volume of water discharged, and hence the resulting temperature.

General benefits seen by the use of VSD PD pumps

The energy that was wasted and returned to the water source was dissipated mainly as heat as it is passed through the regulator. By not pumping at full output this energy wastage is reduced. The alternative would be to run the pump at full flow and use a pressure regulating valve opening to control the feed to the nozzles. The excess flow would then have to be diverted back to source via a pressure regulator, which would require a much larger source tank to ensure that the water temperature rise was acceptable.

As well as the long-term continual energy savings seen by using a VSD pump, rather than wasting energy and dissipating pressure across a valve, the response from a VFD allows improved system controllability.

Appendix A1: Electric motors

A1.1 Energy efficiency

It may well be possible to improve the efficiency of a system by addition of a VFD, however, it will not improve the full load efficiency of a motor. Furthermore, an incompatible motor/drive combination could reduce motor efficiency.

The average electric motor will consume the equivalent of its capital cost in energy in less than two months. Typically, a motor costing €/$ 500 will consume power costing over €/$ 50 000 in its lifetime. This illustrates the importance of giving close attention to efficiency criteria.

Efficiency depends not only on motor design, but also on the types and quantity of materials used. The efficiency can therefore vary considerably from manufacturer to manufacturer.

Manufacturers have focused on the following key factors to improve the efficiency of a motor:

- Electromagnetic design – making the best use of winding and core materials.
- Magnetic steel – utilising a low loss, high permeability electrical steel for the laminations.
- Thermal design – ensuring optimum fit between stator, frame and laminations, and improving cooling paths.
- Aerodynamics – using the most efficient cooling system optimizing fan and fan cover design.

There are several international standards for measuring the efficiency of a motor. European (IEC 60034 and 61972) and North American (IEEE 112) standards vary and will inevitably produce differing results.

In comparing any manufacturers' data, the supply input and test method utilized must be common to each set of data. The most common method in Europe is the IEC 60034-2 "Summation of losses method."

A1.2 Efficiency labelling

A1.2.1 Europe

With the backing of the European Commission, CEMEP representing manufacturers of 80% of the European production of standard motors have agreed to establish three efficiency bands or classes for their 3-phase TEFV (totally enclosed fan ventilated), two and four pole, cage induction motors in the range 1.1 to 90 kW. Most industrial equipment manufacturers and users are therefore affected by this scheme which covers the vast majority of 3-phase motors purchased.

They are arranged in classes: Efficiency 1, Efficiency 2 and Efficiency 3; Efficiency 1 being the most efficient category, and Efficiency 3 the least efficient. The aim is to persuade users to invest in Efficiency 1 equipment.

The efficiency class of each motor, within the scope of the scheme, will be identified in catalogues and on nameplates as a minimum. Some manufacturers will go further and include stickers on the motors for greater visibility.

This will assist equipment manufacturers to demonstrate their more energy efficient equipment to their customers. These labels may, in some cases, be incorporated into other stickers or branding used by the manufacturer to identify a product range. Figure A1.1 illustrates the minimum basic labels. The 'Efficiency' marks are copyright of Gimelec, 11-17, rue Hamelin, 75783, Paris Cedex 16, France.

Motors within the scheme are 400V, 50Hz supply, rated for continuous duty. There are exclusions such as multi-speed motors, special duty motors and flameproof motors.

WIMES 3.03[1] details minimum characteristics for a larger range of motors, beyond those covered by the European efficiency marking scheme. It was prepared by the UK water industry as a basis for specifying the minimum acceptable levels of efficiency in that industry. It is also used as a basis of the UK tax incentive regime for improved efficiency equipment.

The European scheme is voluntary, and in several countries tax advantages or other incentives exist for users of high efficiency motors.

Figure A 1.1: European motor efficiency classification labels

[1] Water Industry Mechanical and Electrical Standard – (WIMES) 3.03 Low voltage electric motors rev 3 Dec 2000, published by the Pump Centre, AEA Technology plc, Birchwood Park, Warrington, Cheshire WA3 6AT, UK. Website www.pumpcentre.com

A Motor Challenge Programme (MCP) has been initiated within Europe and information is available on the EURODEEM website. http://energyefficiency. jrc.cec.eu.int/eurodeem

A1.2.2 North America

Since 1997, all general purpose electric motors manufactured for sale in the United States must comply with the minimum efficiency standards set out in the Energy Policy and Conservation Act, 42 U.S.C 6311 (EPAct) of 1992. The general purpose motors covered by EPAct include the following:

- T-frame, single speed, continuous rated
- Design A or B to NEMA Standard MG1, and IEC frame motors
- Foot mounted (flange mounted motors are not included)
- 230 and 440-480 V 3-phase, 60 Hz, AC Induction motors with cage rotor
- Motors with two, four and six poles

The requirements do not apply to definite or special purpose motors, motors less than 1 hp or more than 200 hp, NEMA design C or D motors, non-NEMA output motors, multi-speed motors and to repaired or rewound motors.

The nominal full-load efficiencies must meet or exceed the values in NEMA Publication MG 1- 1998 (Revision 3), Section 12.58-12.60, Table 12-11.[2]

The legislation has also been adopted by the other members of the North American trading zone, including Canada & Mexico. It applies to motors either sold alone, or as a component of another piece of equipment.

Motors used with frequency converters are included, except for motors with separate cooling.

To identify compliant motors, an approved independent laboratory must undertake the testing and a Compliance Certification number (CC number), which is issued by the Department of Energy, must appear on the motor nameplate, and on all appropriate marketing materials.

Following passage of the EPAct, the US Department of Energy's Motor Challenge program was established to promote increased energy efficiency in motor systems, and the program's activities are now a part of the *BestPractices* activities within the Industrial Technologies program at the US Department of Energy. The *MotorMaster+* software tool available through *BestPractices* is a motor systems energy management tool which simplifies the selection of the best available motor for a given application. *MotorMaster+* and other motor systems-related information are available on the *BestPractices* web site at www.oit.doe.gov/bestpractices.

[2] A condensed version of NEMA MG1 containing all tables referenced in this guide is available for free download from the NEMA website www.nema.org

In 2001, NEMA began a nationwide premium efficiency electric motor program known as NEMA Premium™. The NEMA Premium™ specification includes a minimum efficiency rating beyond the minimum values set by the EPAct, and covers many more motor types and sizes than those covered under the EPAct, including motors up to 500 hp, definite and special purpose motors, and medium voltage motors up to 5 kV. The program is endorsed by many US organizations, including the Hydraulic Institute, Edison Electric Institute, and the Electrical Apparatus Service Association. The Consortium for Energy Efficiency has incorporated NEMA Premium™ into its *Motor Decisions Matter* campaign, which seeks to build awareness about the benefits of motor management and planning. The minimum efficiencies required to meet the NEMA Premium™ efficiencies are detailed in NEMA MG 1-1998 (Revision 3) Sections 12.58-60, Tables 12-12 and 12-13.

More information on the *Motor Decisions Matter* campaign can be found at www.motorsmatter.org. The NEMA Premium™ efficiency electric motor program scope is single-speed, polyphase, 1-500 horsepower, two, four, and six pole, squirrel cage induction motors, For more information about the NEMA Premium™ efficiency electric motor program, go to www.nema.org

A1.3 Motor sizing

Most electric motors are designed to operate continuously at full rated output, at rated voltage. However, it is estimated that only 20% of machines in operation are running at their full rated output. The practice of utilizing a 10% or perhaps 15% margin can often lead to the selection of a higher power rating and, in some cases an increase in the physical size, and therefore cost of the machine.

The loading of the motor will affect its efficiency, Figure A1.2 shows a typical comparison of efficiency between a standard and a high efficiency motor. In most cases the motor will be operating below its rated output. As can be seen in Figure A1.2, the difference in efficiency at full load between the standard and high efficiency motor may be as small as 2–3%, but at half load the difference is considerably greater.

When the decision is made to replace a motor, the user should be aware that some designs of 'high' or 'premium' efficiency motors have a reduced slip characteristic. Such motors run slightly faster, increasing the pump flow, and thus raising the power absorbed by the pump. This factor should be taken into consideration when retrofitting a high efficiency motor onto a centrifugal pump application.

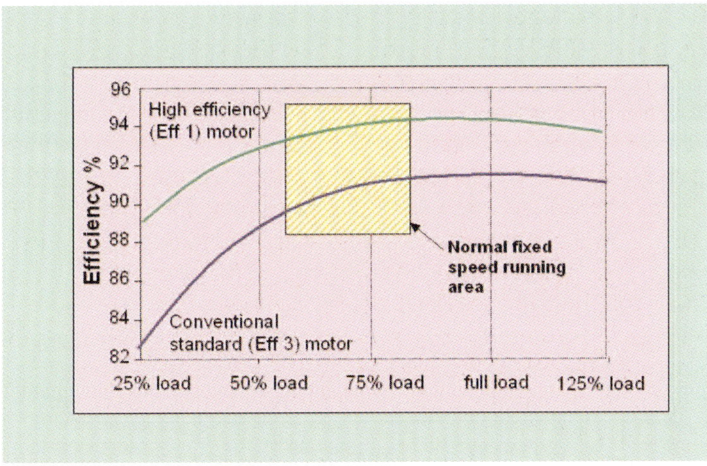

Figure A1.2: Typical full and part load efficiencies for two motors of identical power

Appendix A2: Frequency converters

Chapter 7 lists the majority of frequency converter types. Further explanations of the principal parts of the converter (VFD) i.e. inverter, rectifier and controller are given in this Appendix. Factors to be considered in sizing VFDs are also included.

A2.1 Inverter designs

Table A2.1: Summary of inverter designs and characteristics

Type	Principal Characteristics		
	Advantages	Disadvantages	Availability & Power Range
Voltage Source Pulse Width Modulated PWM	Good power factor throughout speed range. Low distortion of motor current. Wide speed range (2–200%). Multi motor capability. Minimized cogging.	Basic circuit has no regeneration capability. Output pulses may have high rate of voltage change which can affect motor insulation. Output can induce shaft voltages.	Widely available for low and medium voltages. Power range from fractional horsepower to several MW. Many variants available including 4 quadrant low harmonic rectifiers, vector control high dynamic performance, common rectifier feeding multiple inverters.

Table A2.1: Summary of inverter designs and characteristics contd.

Type	Principal Characteristics		
	Advantages	Disadvantages	Availability & Power Range
Voltage-Source Pulse Amplitude Modulated (PAM) Inverter	Simple circuit configuration. Wide speed range (10–200%) Multi-motor capability.	U/f control only Poor power factor at low speeds (unless a diode rectifier and chopper AC/DC converter is used). No regeneration capability. Operation below 10% of rated speed can produce cogging.	No longer commonly available.
Current-Source Inverter (CSI)	Simple and robust circuit design. Inherent regenerative capability. Wide speed range (10–150%).	U/f control only (constant voltage frequency ratio) Bulky. Less efficient than voltage source solutions due to large inductors. Needs tuning to individual motor. Poor power factor at low speed/load. Operation below 10% of rated speed can produce cogging.	No longer commonly available.
Load commutated inverter (LCI)	Similar to CSI.	Often uses several parallel and series circuits for high powers.	Available for medium and high powers (> 1 MW).
Slip Energy Recovery (Kramer)	Simple and robust circuit design. Inherent regenerative capability. Inherent by pass (fixed speed) operation.	Bulky, often needs matching transformer. Requires conventional contactor and resistor starting gear. Less efficient than cage motor solutions due to higher losses in wound rotor motors. Needs tuning to individual motor. Poor power factor at low speed/load. Small speed range (80–98%). Use only with wound rotor induction motor.	No longer commonly available.

Table A2.1: Summary of inverter designs and characteristics contd.

Type	Principal Characteristics		
	Advantages	Disadvantages	Availability & Power Range
Cyclo converter	No intermediate circuit Can operate down to zero speed. High torque capability with field-oriented control. Principle can be used with both induction and synchronous motors.	U/f control only Cannot be used above 33% of input Frequency. Complex circuit design. Poor power factor at low speed. Generates inter harmonics. Can produce power supply notches.	Available for medium and high powers.
Matrix converter	Use with standard cage induction motor. Efficient single conversion without DC link. Inherent regenerative capability.	Requires complex control and powerful microprocessors. Requires new semiconductor layout. Max. output voltage is approx. 70% of input.	Likely future technology
Switched reluctance Drive (SRD)	Used with switched reluctance motor, high efficiency maintained over a wide speed range. Good power factor throughout speed range. High overload torques possible	Depending upon phase count, two multi-core cables may be required between motor and drive. Usually requires feedback from motor mounted encoder.	Power range from fractional horsepower to medium power.

A2.1.1 Voltage source inverter (VSI)

A voltage source inverter controls the output voltage, and the current flowing in the system is limited by the appropriate motor impedance.

VSI converters are available to control AC induction motors, from small (~ 250 W) high dynamic performance servo drives with speed and position control capability, up to most auxiliary drives in industry, ranging up to several thousand kW.

A2.1.2 Voltage source pulse width modulated (PWM) inverter

Given the basic VSI power circuit, the voltage source PWM inverter is now widely accepted as giving the best overall performance.

Details of the design and basic characteristics are given in Section 7.2, Figure 7.4.

Developments in power semiconductor switches have led to the almost universal usage of insulated gate bi-polar power transistors (IGBTs) for low voltage drives, as they are relatively easy to control and switch at very high rates of change, thus minimizing losses in the switch itself.

For low voltage drives it is usual to use a single level DC link, however, at medium voltage it is more usual to use a multi level DC link.

A2.1.3 Voltage source pulse amplitude modulated (PAM) inverter

In a PAM inverter, the DC link voltage is varied either by using a controlled rectifier, which is usually implemented by a six-thyristor bridge, or using a thyristor chopper in the DC link.

The DC is maintained by a capacitor. The motor side converter switches at a single voltage pulse per half cycle, which is sometimes known as a quasi-square wave output. This results in a poorer waveform at the motor terminals in terms of the overall pulse shape, and hence will require substantial motor de-rating, however, it has less effect in terms of voltage stress.

A2.1.4 Current source inverter (CSI)

In current-source inverter (CSI) drives, the inverter switches are fed from a constant current source. Even though a true constant current source can never be a reality, it is reasonably approximated by a controlled rectifier (thyristor or "GTO"), with a current control loop and a large DC link inductor to smooth the current.

Figure A2.1 shows a typical circuit of a CSI. Since the current is constant, there will be a zero voltage drop across the stator winding self-inductance and a constant voltage drop across the winding resistance. Hence, the motor terminal voltage is not set by the inverter but by the motor.

Some manufacturers use modified pulse width modulation techniques to suppress the low frequency 5th and 7th harmonic torque pulsations in the output, especially at low speeds, which are inherent in the six-step

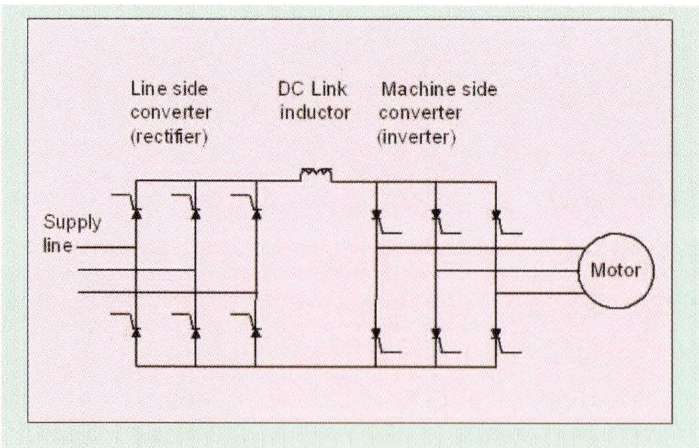

Figure A2.1: Basic CSI main circuit

current waveform. Since the motor parameters must be known to implement such an approach, the CSI drive is presently not popular for general-purpose applications, as it has relatively higher losses than a PWM unit.

A2.1.5 Load commutated inverter (LCI)

The load commutated inverter (LCI), a special type of current-source inverter, is used with very large synchronous motors. The LCI uses thyristors as switching devices but avoids forced commutation because it is used only with synchronous motors.

By controlling the field current, a synchronous motor can run over-excited, that is with a leading power factor, which leads to natural commutation of the thyristors due to the back EMF of the motor. The LCI-synchronous motor combination, although simple and efficient, is generally used only in high power applications due to the higher cost of synchronous motors.

The LCI is also often used as a means of soft starting for large synchronous motors.

A2.1.6 Slip energy recovery (Kramer) drive

Although not strictly a frequency converter, this type of drive is used exclusively with wound rotor induction motors, the functional elements are similar to regenerative converters. The principle of operation is that only the slip energy of the motor, when running, is returned to the network.

This generally involves interfacing transformers to match the rotor voltage with the network.

In operation, the motor is started and accelerated to full speed using conventional starting resistors, and is then switched to the converter, which allows the motor slip to be increased and the speed reduced. The system is only suitable for small speed variations, and the maximum speed is usually lower than for the motor running on fixed speed. The extent of the turn down is related to the rating of the converter.

Slip energy recovery drives are no longer commonly available, as slip ring motors are generally more costly and less efficient than the equivalent cage rotor motor.

A2.1.7 Cycloconverter

This type of drive makes a direct conversion from constant frequency, constant voltage to variable frequency, variable voltage in one stage, without resorting to an intermediate DC link with capacitor for energy storage.

By supplying each phase of the motor winding from a reversible semi-conductor switch, a low frequency AC drive system can be formed. The

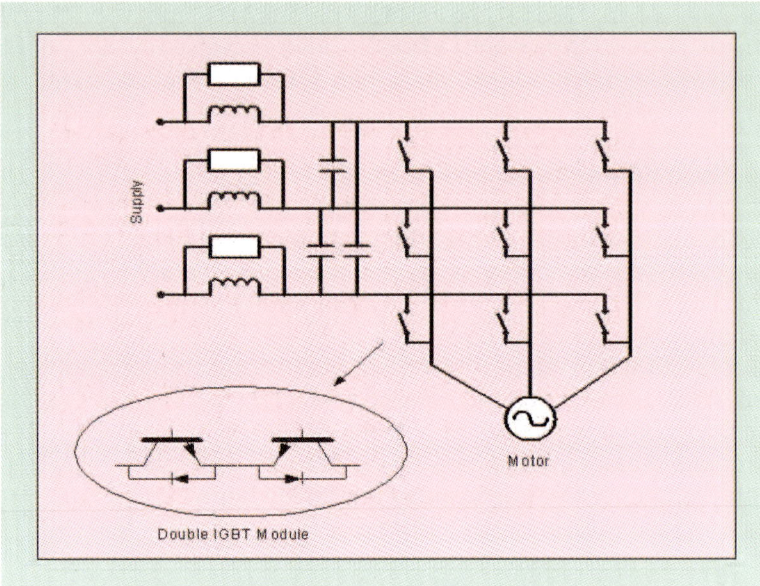

Figure A2.2: Matrix converter main circuit

structure of a cycloconverter involves a large number of power switches and the need for a special three-phase secondary transformer.

Cycloconverters are generally used with high power machines in the MW range for low speed operation (e.g. rolling mills, ship propulsion), although low powered versions are sometimes used for slow speed lifts and hoists.

The maximum output frequency is limited to one third of the nominal network frequency. They can operate down to zero speed and they can be used both with induction and synchronous motors. The main disadvantages are the complex circuit design, the low power factor at low speed, and the harmonic frequency spectrum reflected back to the network.

A2.1.8 Matrix converter

This type of VSD also makes a direct conversion from constant frequency constant voltage to variable frequency/variable voltage in one stage, without an intermediate DC link with capacitor for energy storage. Currently matrix drive technology is in its infancy, but the principle is applicable in all cases where PWM VSI drives are currently used.

The main circuit uses double IGBT modules. 18 IGBT elements with flywheel diodes in nine pairs are required per converter, see Figure A2.2.

The maximum output voltage with a matrix converter is around 70% of the incoming supply voltage, however for many low power applications this is not a particular disadvantage.

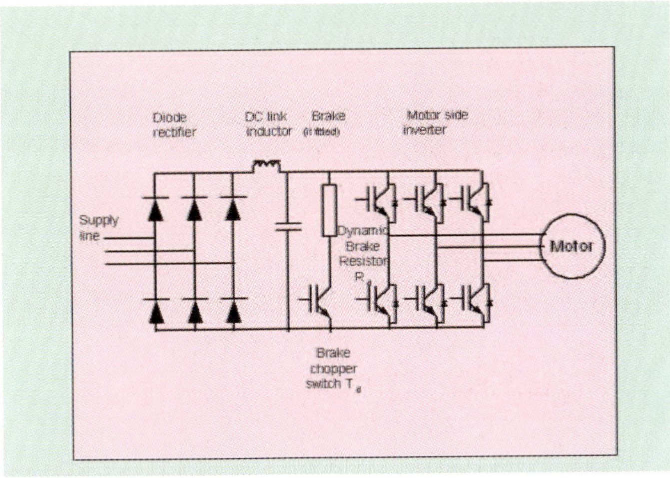

Figure A2.3: Basic PWM drive with dynamic brake

A2.2 Rectifier design

A2.2.1 Dissipative design

The majority of drive systems operate only in the motoring mode, few pumps will be asked to operate in the turbine mode, however, it is an important question that needs to be asked when selecting a suitable drive.

Typically, the generated power from the pump/motor operating in reverse may be dissipated in a dynamic braking resistance in the drive, as long as the power level is low, or when regeneration only occurs occasionally, for instance during reverse running of a deep well pump when the column of fluid empties.

The regenerative power is dissipated in a dynamic braking resistor R_d, that discharges the DC link capacitor when the link voltage rises above a preset value, and triggers switch T_d (see Figure A2.3).

A2.2.2 Regenerative design

When the power rating is large and/or the motor may have to operate for long periods of time in the generating region, the rectifier can be modified to allow regeneration, feeding power back to the supply line.

Present time regenerative drives use a second PWM bridge, to allow that the braking energy be injected back to the source. This feature is used in applications having long time, or continuous operation in the braking mode (ie: in variable speed generation from a water or wind turbine).

Undesirable harmonics can be reduced by choosing a PWM system with sufficiently high switching frequency (limited by the switching devices and potential losses), suitable modulation techniques, and a low pass filter in the line side of the rectifier. The correct selection of the filter also allows enhanced DC link voltages to be achieved.

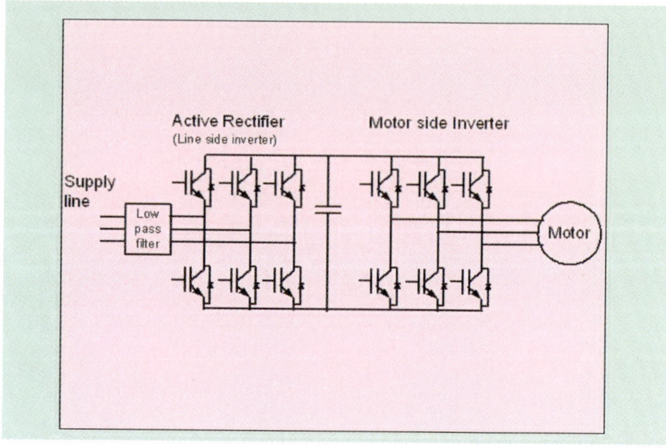

Figure A2.4: Basic active rectifier drive

This is also known as an ***active rectifier*** and allows control of the power factor and harmonic distortion in the supply side. The switching devices used in the AC/DC rectifier and the DC/AC inverter are usually identical (see Figure A2.4).

A2.3 Control strategies

Various measures are used to determine the performance of a drive control system, including the time for recovery from a torque step change or from a speed change signal. In general these measurements are not appropriate in pumping systems where the result of rapid change is generally undesirable. As a general reference in this guide the performance of a closed loop DC motor with a digital control system is considered as a base.

A2.3.1 Scalar or voltage/frequency (U/F) control

Scalar control is applicable to a voltage source inverter, either PAM or PWM.

This is still the most widely used control form for an inverter. The control circuit maintains an approximately constant air-gap flux in the machine by maintaining a constant voltage/frequency (Hz) relationship (for constant shaft torque, the current is approximately constant), as seen in Figure A2.5. Above 100% nominal motor frequency, voltage is held constant and the motor delivers constant power (subject to certain physical and mechanical constraints).

An offset is often added to compensate for the stator resistance voltage drop at low frequency/speed (commonly referred as torque boost or IR compensation). This ensures a fixed level of starting torque, which is adequate for most loads.

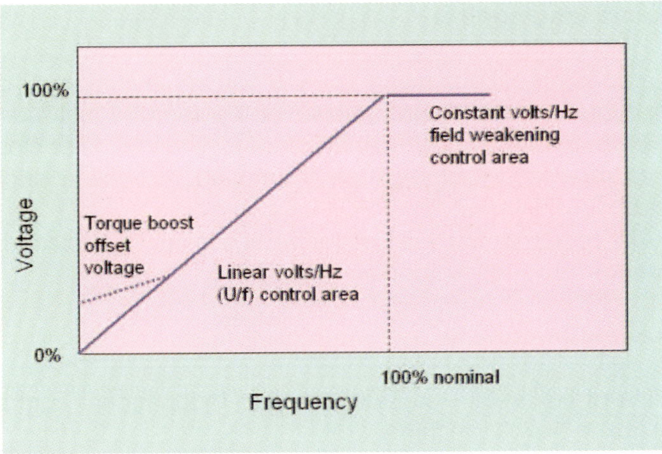

Figure A2.5: Voltage/frequency relation for U/F control

When a step-speed change command is applied, the motor accelerates within a current limit, set by the inverter hardware or control, until a new steady state condition is reached.

The dynamic performance of such systems is relatively poor, with limited torque control capabilities, compared with other control strategies. Speed regulation is limited as no actual speed feedback is available as a control variable, however, because the majority of inverters are used with an external control loop this is not a problem.

Due to its low cost this type of VSD converter control remains the most widely used in energy saving applications e.g. control of fans and rotodynamic pumps in which the speed accuracy is not critical. In these loads, there is no need for high dynamic performance, as long as the speed can be varied with high efficiency over the desired speed range.

The principle is used in PAM inverters by controlling the DC link voltage, and in PWM inverters by varying the pulse pattern. The pulse pattern is controlled by a “modulator”, which holds a suitable algorithm.

Early PWM drives used a combination of a sine wave generator at fundamental frequency and a triangular wave generator, related to the switching frequency, to provide control to the modulator, and are sometimes known as sine/triangle drives.

The majority of currently available drives will have some additional features such as automatic control of the voltage boost, facility for starting onto a spinning load, and some adjustment of the output frequency to maintain steady state speed with an induction motor, generally known as slip compensation.

The overall performance will be generally poorer than a closed loop DC drive.

A2.3.2 Closed loop control

By adding a speed feedback signal, a PDS can achieve higher accuracy in speed. This type of control cannot, however, be used for high performance applications when a wide speed variation (more than 10:1) or close control of torque, is required. This requires further sophistication such as flux vector control.

The overall performance will be generally similar to a closed loop DC drive for speeds >10% of motor base speed.

A2.3.3 Flux vector control

The objective of flux vector control is to give independent control of both torque and flux in an AC machine.

In flux vector control the behaviour of a DC motor is emulated in an induction motor by orienting the stator current with respect to the rotor flux so as to attain independently controlled flux and torque.

Flux vector controllers may also be called ***field-oriented controllers*** and utilise independent control of both magnitude and phase of the AC quantities.

Field-oriented controllers can allow high accuracy speed and torque control in the most demanding applications (e.g. rolling mills, paper machines, hoists and lifts).

The modeling used in flux vector control is relatively complex and for optimum performance the actual rotor position is required, by closed loop control, using a shaft mounted transducer. The output waveform of a flux vector controlled drive is set by the PWM modulator.

Flux vector control will allow automatic control of the voltage boost required for starting and provide a substantially higher starting torque, than available from scalar control.

The overall performance will be generally equal to or better than a closed loop DC drive for speeds >5% of motor base speed, with all drives, and from zero speed with closed loop drives.

A2.3.4 Direct flux and motor torque control

Direct flux control or direct torque control (DTC) are other implementations of a field-oriented controller and use a sophisticated motor model.

Typically in DTC, the processor will set the maximum number of switching events, but the actual initiation of the switch positions is undertaken by the control processor, eliminating the modulator. This is sometimes described as "just in time" switching.

DTC also allows very high levels of motor starting torque to be achieved, limited only by the current output capability of the inverter and the physical construction of the motor.

The overall performance will be generally equal to or better than a closed loop DC drive for speeds more than twice the motor slip speed, with all drives, and from zero speed with closed loop drives.

A2.4 Factors to be considered in sizing variable frequency drives

This Chapter gives an outline of the factors to be considered in the sizing of a VFD for more straightforward applications. Other parameters such as environmental conditions, harmonic considerations, multiple pump applications, load inertia, need for braking, control capabilities and communication may also affect the selection of the VFD. It is recommended in all cases to consult the VFD manufacturer for an engineered solution for the specific application.

Four main considerations come into play when sizing a VFD pumping application:

A2.4.1 Motor nameplate full load amps (FLA)

The motor full load current is the most important factor in sizing a VFD and for electronic motor protection. The motor power rating (kW or hp) value may be used only as an approximation as the power factor and efficiency of the motor can vary substantially, especially for slow speed and electro submersible pump motors. The VFD size will normally be selected with a maximum continuous output current rating that is greater than the motor full load current. It is not required to grossly oversize a VFD for a pumping application unless motor uprating is foreseen in the future.

If the duty is well specified, and the actual maximum running current is known then there are possible cost saving opportunities when retrofitting a constant speed application with a VFD. This is achieved by selecting a drive with lower rated current than the motor nameplate. Case Study 1 provides an example, but this selection must be undertaken with considerable care.

A2.4.2 Motor load torque characteristic

VFD manufacturers will often allocate several ratings to a particular converter. For example, a rating with small overload capability, typically 10% or 20% for a short duration, (often one minute in ten), and also a lower rating, with rather higher short-duration overload capability of 150% or 200%. The purpose of these overloads is to allow short period transients required for acceleration and overcoming starting stiction.

The very different torque-speed curves for rotodynamic and positive displacement pumps are shown in Figure 7.3. These two load characteristic are sometimes referred to as variable torque and constant torque loads respectively. In general, a rotodynamic pump can be powered by a VFD with small overload (overcurrent) capability, whereas a PD pump may well require a higher overload capability to achieve starting requirements.

In general, if a VFD is rated for variable torque loads it will have a higher maximum continuous current output than with the same VFD rated for a constant torque application. Conversely, if a VFD is rated for variable torque loads it will have a lower percentage of over current capacity than the same VFD rated for a constant torque application.

A2.4.3 Motor nameplate 3-phase voltage

It is always important to match the voltage that can be fed from the VFD to the motor being used. A VFD will be quoted as having an input supply tolerance of +/–10%, but in general it cannot produce more volts than it is supplied with. In fact, generally the maximum output voltage from a VFD will be lower than the supply volts, due to volt drops across the inductors, used to reduce harmonics, and in output line (dU/dt) filters. (Note: Active rectifier drives, see Section A2.2.2, are capable of boosting voltage.)

In some cases step up transformers may be appropriate, but the advice of the motor and VFD manufacturer should be obtained, as these transformers may require a sine filter in the inverter output, and a special core construction.

If the voltages are mismatched, the VFD may have to pass a higher current to the motor to achieve a required torque, with consequent additional heating of the motor.

A2.4.4 Single phasing

In rural areas, three phase power is not always available. Generally single phase supply VFDs are limited to powers up to around 2.2 kW (3 hp).

Some VFD manufacturers will allow their standard three phase VFDs to be supplied at single phase, providing they are suitably de-rated, (typically by a factor of two), and with the phase loss protection inhibited.

Some manufacturers offer VFDs for higher power single-phase applications. Proprietary series harmonic filters are available which will provide a three phase supply, with reduced net harmonics, allowing a conventional VFD to operate off a single phase supply.

While single phase supplies are used to feed the VFD, the output to the motor is always three phase, allowing lower cost three phase motors to be used. Single phase capacitor started motors must never be used with a VSD.

The cost of these solutions must be weighed against bringing a three phase power to site.

Appendix A3: Legal obligations and harmonic suppression

A3.1 European Directives

The interrelationship of a system with its environment is laid down by a number of EU Directives and enforced by national legislation. These are direct obligations on manufacturers, users and their agents and must be applied when installing and operating new or upgraded equipment.

A number of EU Directives are particularly important when considering variable speed pumping applications.

A3.1.1 The Machinery Directive[3]

This directive applies to all combinations of mechanically joined components, where at least one part is moving and which have the necessary control equipment and power input circuits.

Compliance with the Directive is achieved by implementing the requirements of a number of harmonized standards.

From the point of view of an electrically driven pump, the most important of these standards are likely to be EN 60204-1 & EN 809.

A3.1.2 The EMC Directive[4]

Electromagnetic compatibility (EMC) is the ability of a device or system to function without error in its intended electromagnetic environment and also to operate without emitting interference, which would disturb other devices in that environment.

[3] Directive 98/37/EC published by the EU Commission, Rue de la Loi 200, B-1049, Brussels.
[4] Directive 89/336/EEC, modified by 91/263/EEC and 93/68/EEC published by the EU Commission, Rue de la Loi 200, B-1049, Brussels.

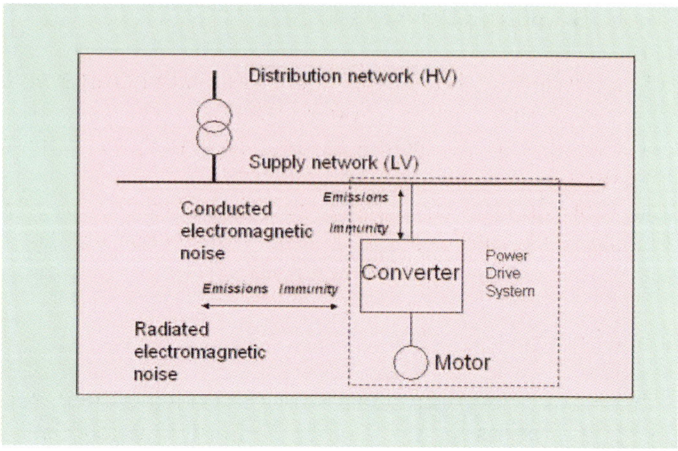

Figure A3.1: EMC aspects of a Power Drive System

The EMC Directive details the levels of electrical disturbances, both radiated and conducted, that the drive system is permitted to emit and also the levels to which it must be immune. The types of emissions are shown schematically in Figure A3.1

Radiated emissions – The drive is a source of high frequencies due to the controlling microprocessors and the rapidly rising wavefronts of the output voltage pulses. Unsuitable cabling and incorrect grounding can allow these high frequencies to radiate and disturb the environment.

Conducted emissions **–** Lower frequency disturbances can be passed back through the supply network to cause problems for other users of that network. To meet the requirements of the Directive, a power drive system must comply with the harmonised Product Specific Standard EN 61800-3.

A low voltage (<1000 V) first environment point of connection has more stringent requirements than a high voltage (> 1000 V) second environment connection.

There is also a division by the manner in which the equipment is placed on the market, either by unrestricted distribution where the vendor has little or no EMC expertise, or by restricted distribution.

To avoid problems, it is essential that the manufacturers' installation recommendations are strictly followed, including:

- Correct selection of main and control cables
- Proper grounding
- Proper shielding
- Installation of suitable filters for conducted disturbances
- Isolation transformers with earthed screens between the windings
- For submersible pumps with unscreened down well cables a suitable output filter may be required

For further details of control of harmonics see Section A3.3

A3.1.3 The Low Voltage Directive[5]

This Directive concerns all electrical equipment with nominal voltages from 50 V to 1000 V AC (and 75 V to 1500 V DC).

The aim is to provide protection against electrical, mechanical, fire and radiation hazards.

The Low Voltage Directive does not apply to products for use in potentially explosive atmospheres, these are covered by the ATEX Directive.

A3.1.4 The Atex Directives[6]

These Directives provide a framework for determining the essential health and safety requirements for use of any product in a potentially explosive atmosphere.

There are two areas covered, one is product related and one is related to the essential safety of the worker on site.

The Directives apply to the entire equipment INCLUDING the pump. They also include operation within dust hazards, as well as flammable gases and replace the Low Voltage Directive for equipment installed in potentially explosive atmospheres.

In general, the installation will require a de-rating of the motor and additional protection to be installed.

Under the terms of the Directive (94/9/EC) it is unlikely that any electric motor manufactured under previous standards, or meeting earlier versions of this Directive will be permitted to be retrofitted with a variable speed drive.

When considering an installation in a hazardous area the advice of the motor and converter manufacturer must be sought to enable a properly matched and certified system to be achieved.

A3.1.5 The CE Marking Directive[7]

Under the terms of the EU Directives, it is a statutory requirement that all mechanical, electrical and electronic systems, installed in Europe, meet the requirements of all appropriate Directives, and carry the appropriate "CE" compliance indication.

For all equipment covered by EU Directives the principal document showing conformity is the Manufacturers' Declaration of Conformity. The pumping system should not be commissioned without the appropriate declarations to the EMC, Low Voltage and Machinery Directives, plus the ATEX Directive if applicable, being available.

[5] Directive 72/23/EEC, modified by 93/68/EEC published by the EU Commission, Rue de la Loi 200, B-1049, Brussels.

[6] Directive 94/9/EC (Product Directive) and Directive 1999/92/EC (Worker Protection Directive), both published by the EU Commission, Rue de la Loi 200, B-1049, Brussels.

[7] Directive 93/68/EEC published by the EU Commission, Rue de la Loi 200, B-1049, Brussels.

A3.1.6 Environmental legislation

The EU is currently considering a number of measures which are looking at directing manufacturers into a full life cycle integrated product policy to reduce environmental impact.

Directives have already been initiated on the topics of hazardous waste, use of lead, noise, packaging, recycling and disposal (waste electrical and electronic equipment) and energy using products.

Further work is planned on energy demand management, energy efficiency requirements for end use equipment and the environmental impact of electrical equipment.

A3.2 United States regulations and standards

A3.2.1 Federal regulations

In the United States, one federal regulation regarding electromagnetic compatibility applies to Variable Frequency Drives. Part 15 Subpart B of Title 47, Volume 1, Chapter 1 of the Code of Federal Regulations addresses high frequency emissions from “unintentional” radiators. All digital devices (including drives) which are not used exclusively in a public utility or industrial plant must comply with the high-frequency radiated and conducted emission levels required in Subpart B. Conducted emission limits required by Subpart B are identical to those of CISPR 11 Group 1 limits, and Subpart B accepts the radiated emission limits of CISPR 11 in lieu of those required by Subpart B.

A3.2.2 Safety regulations and standards

Other safety standards applicable to VSDs, for which compliance is legally either voluntary or elective in nature, may be required by authorities in civil positions, such as county, municipal, state or federal governments. When applicable, these requirements are usually identified in bidding documents. Some of these are described below.

UL 508C Power Conversion Equipment from Underwriters Laboratory (UL) is used by many in the US to qualify electrical and thermal safety of VSDs, and can be required by municipal or other authorities.

NEC (National Electric Code) also known as *NFPA 70* from the National Fire Protection Association.

NFPA 79 Electrical Standard for Industrial Machinery from the National Fire Protection Association.

OSHA (Occupational Safety and Health Administration of the US Govt.) Regulations (Standards – 29 CFR).

In addition, NEMA ICS 7 and ICS 7.1 from the National Electrical Manufacturer’s Association address safety practices for adjustable speed drives. Consideration is currently being given for these documents to be replaced by IEC standards IEC 61800-1 (DC drives) and IEC 61800-2 (AC drives).

A3.2.3 Harmonic current emissions

There are no government standards setting limits for harmonic current emissions. However, some customers and power utilities in the US have chosen to impose either their own standards or have adopted *IEEE 519 Recommended Practices and Requirements for Harmonic Control in Electrical Power Systems.* The techniques described in Section A3.3 are among those that IEEE 519 encompasses. IEEE has also published a document useful in the application of harmonic limits, *IEEE 519A Guide for Applying Harmonic limits on Power Systems.*

A3.3 Harmonics

This Chapter considers in more detail low frequency emissions, or harmonics, and methods of attenuation. The level of harmonics reflected back to the supply network is usually regulated by the electricity supply utility.

Harmonics are voltages and currents in the electrical system at frequencies that are multiples of the fundamental frequency: 50 Hz in European power systems, 60 Hz in USA and certain other countries.

Harmonics are associated with any load that uses a rectifier based power supply such as radio and TV, computers, lighting ballasts, other domestic equipment such as washing machines and microwaves. Harmonics also come from generators and transmission equipment. They can affect the equipment performance and are both caused by and can interfere with the function of VFDs.

Harmonics increase equipment energy losses and may create excessive currents and heating in transformers and neutral conductors. Harmonic waveforms are characterized by their amplitude and harmonic number. The illustration in Figure A3.2 shows the distorting effects of a number of harmonic currents added to a fundamental frequency.

It can be seen therefore that any rectifier will pass harmonic currents back to the network, where they will generate voltages, when applied to the supply impedances.

Within Europe, legislation exists to limit current harmonics for devices up to 16 A under the EMC Directive, and supply utilities also have a duty to regulate the harmonic voltage distortion levels in their networks to prevent disturbance with other users. An example of these regulations is the British Electricity Association Engineering Recommendation G5/4.

In the United States the IEEE Standard 519 is generally applied, as explained in Section A3.2.

Sophisticated control techniques have been developed in modern PWM type VFDs to minimize the effects of harmonics on the motor side. Various methods are also available to reduce the harmonics reflected back to the network.

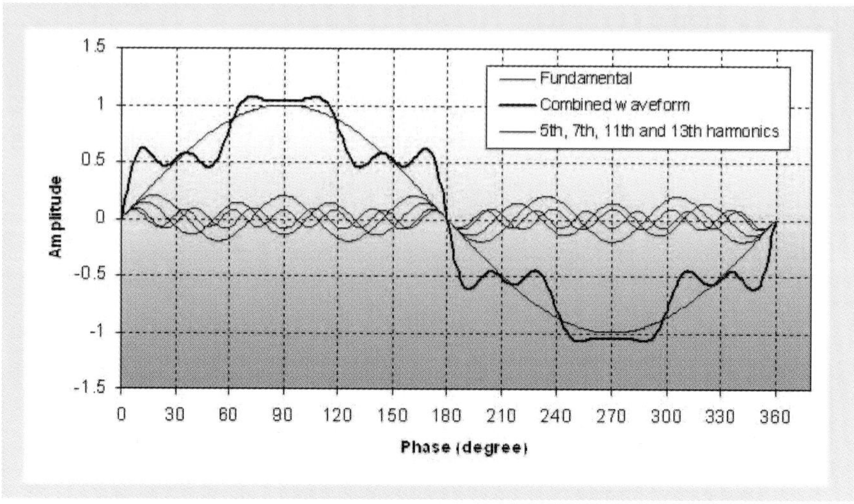

Figure A3.2: Typical waveform with characteristic six-pulse VFD load[8]

Techniques include:–

- Adding input or DC link chokes
- Utilizing active or passive filters
- Using multi pulse rectifiers with associated phase shifting transformers
- Using active rectifiers

The techniques are detailed in the following Chapters, with a summary in Table A3.1 and a comparison of residual harmonics in Table A3.2

A3.3.1 Adding input (AC line) or DC link chokes

The most common rectifier circuit used in a three-phase VFD is the six-pulse diode bridge. It consists of six uncontrolled rectifiers or diodes and a DC capacitor to maintain the DC link voltage. A choke, can be installed either in the AC supply or in the DC link, which will work together with the DC capacitor to form a low-pass filter which smoothes the DC current.

The choke may be referred to as an inductor or a reactor (see Figure A3.3).

Chokes are a simple and cost effective solution for harmonic suppression, but may not be effective enough against low order (5th and 7th) harmonics in multiple installations or in medium and high power applications.

A3.3.2 Passive filters

Passive filters are generally tuned devices designed to remove specific harmonic currents from a source. All passive filters are network sensitive, and require care in application.

[8] EU SAVE Study on VSDs for Electric Motor systems

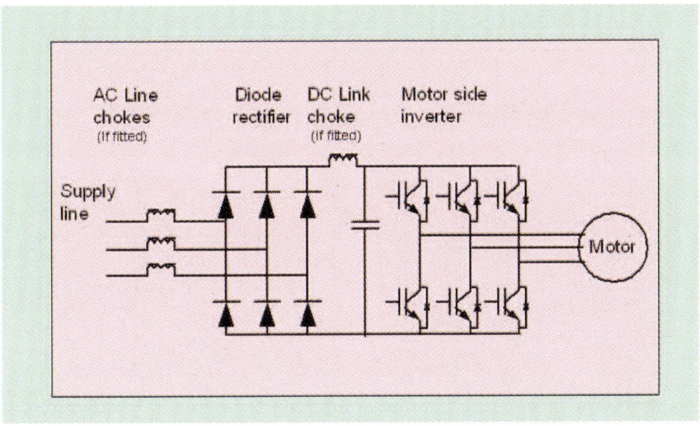

Figure A3.3: Basic PWM drive with optional chokes

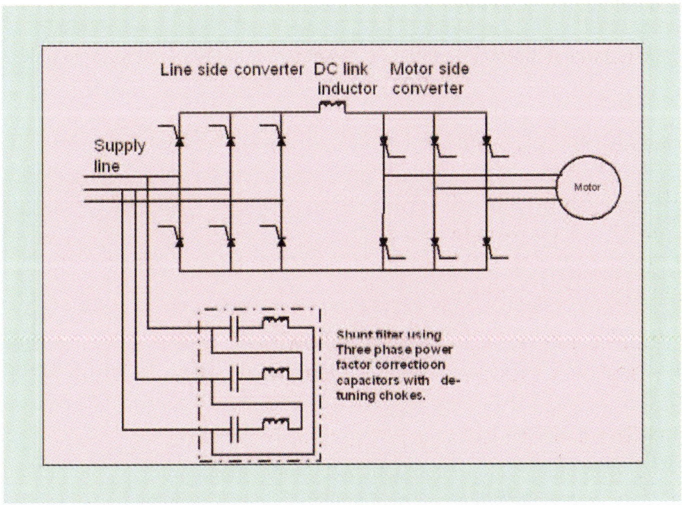

Figure A3.4: VFD with shunt filter

A3.3.2.1 Shunt filters Shunt (or parallel) filters are a traditional way of removing harmonics, but can only be effective when correctly tuned. Figure A3.4 is a very simplistic diagram of the circuit.

Each filter is designed to remove a specific harmonic, thus multiple filters may be required in some installations. Passive filters incorporate capacitors which will correct reactive power, thus they can be useful for DC drives and for CSI and LCI converters, but can be harmful with diode rectifier drives, which will not operate satisfactorily with a leading power factor.

A3.3.2.2 Series filters By connecting a number of inductors and capacitors in series with the input to a drive, it is possible to create a higher impedance path for harmonics than for the fundamental current (see Figure A3.5).

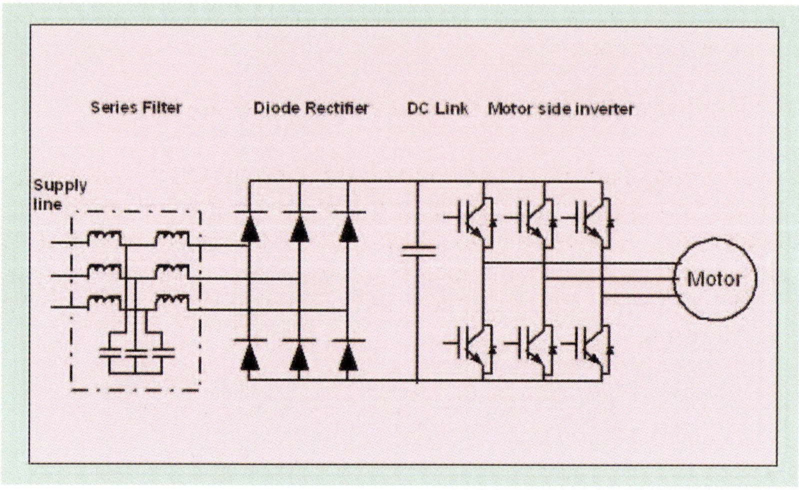

Figure A3.5: VFD with series filter

Series filters can resonate with the DC link capacitors in a drive, and are generally suitable for systems with relatively slow dynamic responses. They also tend to be quite large and have higher losses associated with them compared to active solutions, described in Section A3.3.3.

A3.3.3 Active harmonic filter

The active filter uses a PWM inverter, with associated capacitors and inductors. It is designed to inject harmonic currents in anti-phase to those produced by other equipment on a network, and hence reduce overall distortion (see Figure A3.6).

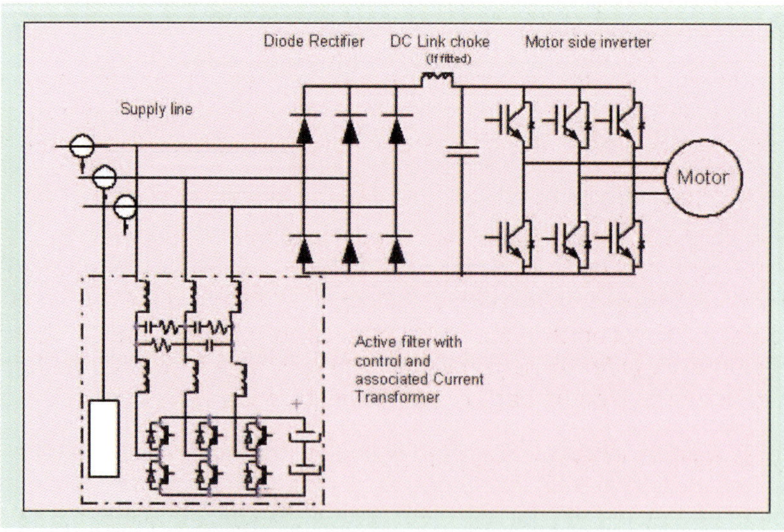

Figure A3.6: VFD with active filter

Active filters are more suited for multiple small VFD applications due to their relatively high cost.

A3.3.4 Multipulse rectifier with a phase shifted transformer

On the secondary side of a three winding transformer the Δ and Y outputs are 30° out of phase causing certain of the of the harmonics generated on the supply side of each rectifier to be opposite and thus cancel out. This is a 12-pulse system as it comprises 2 x six-pulse rectifiers (see Figure A3.7). This solution may be extended, using suitable transformer windings to 18 or 24 pulses.

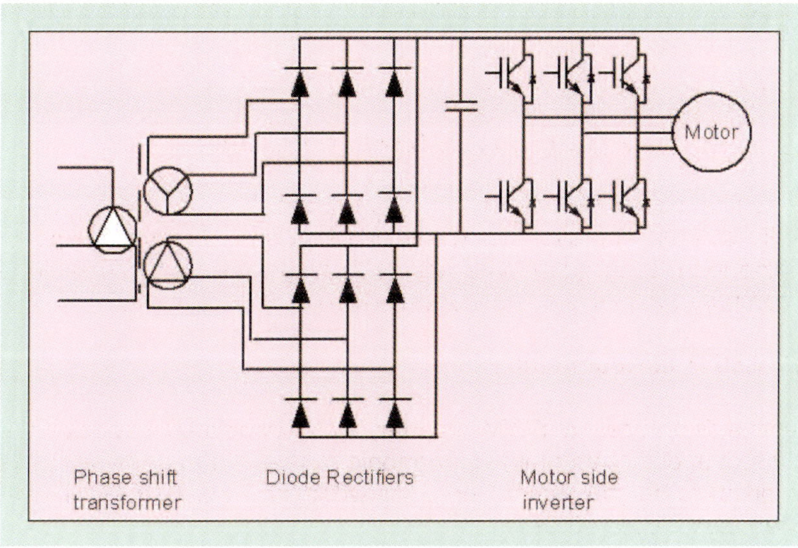

Figure A3.7: 12-Pulse rectifier

When two pumps are designed to run at the same time with the same load, a cost-effective solution may be to use two VFDs with six-pulse configuration in conjunction with a phase shifted transformer. This becomes a quasi 12-pulse drive.

This design will produce comparable harmonics cancellation to the 12-pulse design provided the loads remain approximately the same.

Another advantage of using phase shifted transformers is that higher secondary voltages (up to 690 V AC for IGBT based VFDs) may be selected to reduce current thus reducing inverter modules and cable size. This design is ideal for large pumping systems where a power transformer needs to be purchased as part of the new installation.

A3.3.5 Active rectifier (four-quadrant VFD)

Four-quadrant PWM VFDs use a line side inverter module as an active rectifier, as discussed in Section A2.2.2.

The active rectifier enables two-way energy flow; no dynamic brake resistor circuit is required when braking, the excess energy is sent back on the input line. four-quadrant VFDs using active rectifiers do not generate any significant levels of harmonics but can cost between two and three times more than six-pulse rectifier VFDs (see Figure A3.8).

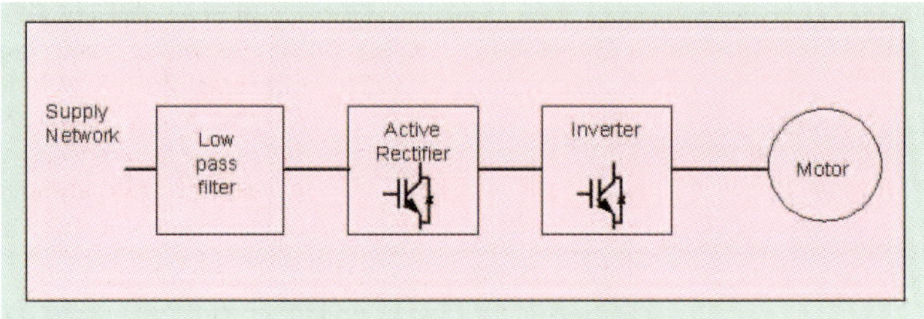

Figure A3.8: Active rectifier

This extra cost may be justified on high inertial dynamic braking applications (not pumping typically) and on installations where power transformers already exist.

A3.3.6 General

The use of harmonic attenuation measures with passive filter and with phase shift transformers must be carefully considered where the site is likely to have standby (emergency) generators, where the effective supply impedance can change substantially. The attenuation systems and their relative effectiveness are summarized in Tables A3.1 and A3.2.

Table A3.1: Summary of harmonic attenuation systems

Technique	Advantages	Disadvantages
Adding input line chokes or DC link chokes	Simple and cost effective for improvement to high frequencies.	Volt drop across chokes can limit output voltage to motor.
Shunt passive filter	Cost effective solution if reactive power available for compensation.	Network sensitive, changes in net parameters can require retuning. Can result in leading power factor giving diode rectifier miscommutation, and capacitor damage.
Series passive filter	Cost effective for single, medium size drive.	Large physical size. Tends to limit total current distortion, residual usually all 5th harmonic.
Active harmonic filter	Versatile – can select specific harmonics, and filter to limits. Can be retrofitted to existing network.	Relatively costly.
12-Pulse rectifier. Multipulse rectifier with a phase shifted transformer	Simple passive solution. Best suited to applications with dedicated transformer.	Requires special transformer. Higher losses, especially if fed from low voltage network. Large physical size.
Active rectifier (four-quadrant VFD)	Simple solution. Operates well with low fault levels. Can operate in generator mode.	Relatively costly.

Table A3.2: Typical data for harmonic attenuation systems

Technique	Relative physical size	Relative cost	Typical THD current %	Typical harmonic current content				
				Funda-mental %	5th %	7th %	11th %	13th %
No attenuation six-pulse diode rectifier Shown for reference	100	100	71	100	60	36	8.3	8.9
six-pulse diode with AC choke diode rectifier	120	120	40	100	36	14.2	7.8	4.3
six-pulse diode with DC choke diode rectifier	115	120	36	100	32	12	8.9	5.6
six-pulse diode with series filter	150	160	8	100	7.9	1	0.1	0.1
12-pulse diode rectifier	210	300	10.6	100	2.5	1.6	8.5	5.4
Active rectifier	220	210	1	100	0.6	0.4	0.5	0.4

THD = Total Harmonic Distortion
Other harmonics are not normally significant
Filters are excluded as the levels of filtration will depend on the specific installation.

Appendix A4: Frequently asked questions

A4.1 Motor protection

Q – For most motors the thermal loadability drops as the speed reduces, how do we protect the motor?

A – A conventional motor thermal overcurrent relay would allow a motor to draw a too high current at low speed, and is therefore unreliable.

An embedded thermal device is one method of obtaining protection. Thermostats, PTC thermistors or RTD devices may all be used with an appropriate external relay some drives will also have integrated thermistor monitoring.

As the inverter controls the speed and monitors the motor current, it can also provide motor protection, by comparing the load, speed and loadability.

Some more sophisticated drives offer both over and under load protection, which allows monitoring of coupling failure or possibly low flow, in a rotodynamic pump. It should also be noted that, in many rotodynamic pumping applications with a typical variable torque load curve, where the load is reduced with the square of the speed reduction, the resultant motor heating is lowered and overheating should not be a problem.

A4.2 Speed

Q – What limits a motor's maximum speed?

A – A motor is not limited to its nominal speed determined by the mains supply frequency. It can be both reduced or increased. Figure 10.1 shows that beyond base speed the torque available will drop as the inverter cannot increase its output voltage, the motor becomes progressively under fluxed, this is known as field weakening.

There is a specific maximum output speed for any motor, based on two factors:

Firstly the mechanical maximum speed, which is normally limited by the bearings and peripheral velocities, and secondly the electrical maximum speed, which is limited by the reduction in flux, causing the available torque to be reduced. Typical values of maximum speed for standard TEFC motors with grease lubricated ball bearings are shown in Table A4.1.

Table A4.1: Typical maximum speeds for TEFC motors

IEC Frame Size	NEMA Frame Size	Max speed r/min
63 – 100L	48 – 145	6000
112 – 200L	182T – 326T	4500
225 – 280H	364T – 449T	3600
315 (2 pole)	505 (5005) – 5011	3600
315 (4, 6, 8 pole & slower)	505 (5005) – 5013	2400
355 – 400 (2 pole)	586 (5806) – 5812	3600
355 – 400 (4, 6, 8 pole & slower)	586 (5806) – 5812	2400

Maximum speeds for NEMA motors are detailed in specification NEMA MG1 Tables 12.5 and 12.6.

For larger two-pole motors it is common to design with a “flexible shaft”, where a mechanical resonance may be present below the design running speed, which also precludes running at critical speeds below the nominal speed.

The advice of the manufacturer should be sought for motors typically above 400 kW (500 hp) especially before:

- Running above its base speed
- Running a two-pole motor at variable speed
- Running motors which include self lubricated sleeve bearings at variable speed

Motors specifically designed for higher speeds are becoming more readily available.

A4.3 Insulation design

Q – The press has carried many scare stories about insulation failure with inverter fed motors, what standards are required?

A – Modern motor insulation systems are suitable for inverter use in Europe based on a 400 V nominal supply, however, always seek the manufacturers' confirmation. Higher supply voltages (440 to 1000 V) including those common in the US and Canada will normally require "inverter rated" windings. These are specified in NEMA standard MG1 pt 31 and many manufacturers now offer these as standard. [9]

In addition, remedial measures such as fitting filters in the inverter can often be taken to permit connection of lower rated equipment.

For high voltage motors, the advice of the drive and motor manufacturer is always recommended. Many high voltage drives incorporate some degree of output filtering to relieve insulation stress.

A4.4 Motor bearings

Q – The press has carried even more scare stories about bearing failure with inverter fed motors, what actions are required?

A – There are two points that need to be taken into account. Motor bearings may run hotter with an inverter fed motor, as some additional losses will occur in the rotor, and are dissipated down the shaft, potentially requiring high temperature lubricants, or more frequent lubrication.

In addition, with physically larger motors, it is possible for voltages to be induced in the shafts resulting in circulating currents through the bearings (this phenomenon has actually been known for nearly 100 years). To reduce the impact, it is recommended to fit insulated bearings at the none drive end (NDE) of all motors rated over 90 kW (125 hp) or with shaft centreline heights of over 280 mm (11 inch).

Good cabling and grounding practice is also critical to reducing the levels of induced voltage.

In addition some common mode filtering may be needed with very large motors, or on motors fed at 690 V or greater.[10]

A4.5 Sleeve (journal) bearings

Q – What are the effects of a VSD when equipment is fitted with sleeve (journal) bearings?

A – There are several points that need to be taken into account. Many motors and pumps are designed with self lubricated sleeve bearings. In these, a flinger ring ensures that oil is displaced into the bearing during rotation. The amount of oil transferred is dependent on speed. Other bearings utilise a pump fed system that provides a constant oil volume.

---

[9] REMA/GAMBICA Technical Report No 1 "Motor Insulation Voltage Stresses under PWM Inverter Operation.
NEMA Standards Publication – Application Guide For AC Adjustable Speed Drive Systems
[10] GAMBICA/REMA User Guide No 2 "Motor Shaft Voltages and Bearing Currents"

In either case the pump manufacturer should be consulted as to whether the bearing is suitable, and as to whether a self lubricated bearing can be converted into a force lubricated system.

With a multistage pump, there are also inter-stage bearings, and the manufacturer again must be consulted.

A4.6 Reliability

Q – What is the likely effect of including a drive within my system?

A – Frequency converters have been available in the market place for over 25 years, and have improved in reliability over this time.

Subject to satisfactory installation and routine maintenance, including removal of dust, cleaning cooling surfaces, and regular replacement of limited lifetime parts, such as cooling fans, together with regular pump and motor bearing lubrication and maintenance, then a modern drive system should have a long and satisfactory life.

Pump and motor reliability will generally improve because the reduced running speed, for at least part of the duty cycle, will reduce loads on the equipment.

Current manufacturers' experience shows the MTBF for the electronics increasing. Currently it is around 15 years to 30 years, depending on size and installation.

Appendix A5: Effects of pump speed and impeller diameter on magnetically driven pumps

A5.1 Changes of pump speed

All pumps are designed to move liquid and increase pressure, but to fulfill this requirement the liquid must be kept safely within the pump pressure casing and away from the atmosphere surrounding the pump.

There are two main types of pump sealing design for rotodynamic pumps that are used to achieve this function. These react differently to speed change.

These two pump sealing designs are:

- The impeller is driven directly, with the shaft sealed by gland packing or mechanical seal where it enters the pump casing.
- Pumps sealed with a stationary device called a containment shell, sheath or shroud. With this design, the pump shaft is not driven mechanically by the motor, but is driven instead by magnetism across the stationary containment shell, either by the use of permanent magnets or an eddy current drive.

For synchronous magnet driven pumps the technique for calculating the change in total pump power with speed is shown below.

Section 4.2.1.1 describes the affinity laws, which apply to the hydraulic performance of pumps i.e.:

$$Q \propto n$$
$$H \propto n^2$$
$$P \propto n^3$$

Where:

Q	=	Flow rate
H	=	Head
P	=	Power absorbed
n	=	rotational speed

However, with magnetically driven pumps, as well as hydraulic power, there are usually magnetic power losses to be taken into account. These are proportional to speed squared. (The exception to this rule, is that there are generally no magnetic losses incurred for non-metallic magnetically driven pumps.)

For a magnetically driven pump the total power absorbed by the pump is the sum of hydraulic power and magnetic losses.

So, total power = hydraulic power (P) + magnetic losses (M)

hydraulic power follows the affinity laws, i.e. $P_2 = P_1 \times (\text{r/min}_2/\ \text{r/min}_1)^3$

magnetic power (M) follows the physical law:

$$M_2 = M_1 \times (\text{r/min}_2/\ \text{r/min}_1)^2$$

Additional reading is available.[11]

The following example illustrates the overall effect on power for a magnetically driven pump, for a required speed change from 2900 r/min to 1650 r/min:

The total power absorbed by the pump is 110 kW at 2900 r/min.

The pump manufacturer has advised that the magnetic coupling losses (M_1) are 10 kW.

So the hydraulic power part (P_1) is 110 kW – 10 kW = 100 kW.

Using the affinity laws,

$$P_2 = P_1 \times (\text{r/min}_2/\ \text{r/min}_1)^3$$
$$P_2 = 100 \times (1650/2900)^3 = 18.4 \text{ kW.}$$

Magnetic losses change to

$$M_2 = M_1 \times (\text{r/min}_2/\ \text{r/min}_1)^2$$
$$M_2 = 10 \times (1650/2900)^2 = 3.2 \text{ kW.}$$

So the total power at the new speed of 1650 r/min = 18.4 + 3.2 = 21.6 kW.

Converting the total power just using the affinity laws would have predicted a power at 1650 r/min of $110 \times (1650/2900)^3 = 20.2$ kW, which is an error of nearly 7% in this example.

This could result in an overloaded motor.

Cautionary Notes:

1. The above calculations apply to the conversion of pump performance curves. In an actual application the system curve must

[11] Referenced reading on this effect is Para 1.6.5.8.8.1 in ANSI/HI-1.6-1994 Hydraulic Institute – Centrifugal Pump Tests.

be taken into account to determine the actual operating points and associated power absorbed.

2. For any case where there is a significant increase in speed, always verify with the pump manufacturer that the magnetic coupling is adequately rated for this new power and speed.
3. Non-metallic magnetic drive pumps usually have no magnetic losses and so power can be calculated by the affinity laws, similar to mechanically sealed pumps. However, the pump manufacturer must still always be consulted to check that the magnetic coupling is adequately rated for the new power, speed and generated pressure, if speed is increased.
4. For other types of magnet drive pump, such as torque ring or induced drives, which can be used for very high temperatures, the calculation of power change with speed is more complex and the manufacturer should be consulted.

A5.2 Change of impeller diameter

From Section 4.2.1.2 the equations defining the variation of pump hydraulic performance with impeller diameter *D* are:

$$Q \propto D$$
$$H \propto D^2$$
$$P \propto D^3$$

For a magnetically driven pump, the effect of diameter change on power, only applies to the hydraulic power. If the speed is unchanged the magnetic losses will not change. Using the above example for a diameter change from 250 mm to 220 mm

The total power absorbed by the pump is 21.6 kW at 1650 r/min.

The pump manufacturer has advised that the magnetic coupling losses are 3.2 kW.

So the hydraulic power part is 21.6 kW – 3.2 kW = 18.4 kW.

Using the affinity laws, $P_2 = P_1 \times (D_2/D_1)^3$

$$P_2 = 18.4 \times (220/250)^3 = 12.5 \text{ kW.}$$

Magnetic losses are not affected by a change in diameter so remain at 3.2 KW

So the total power at the new diameter of 220 mm = 12.5 + 3.2 = 15.7 kW.

Appendix A6: Efficiency of variable speed drive systems

A6.1 Energy efficiency in the supply chain

Each step in the supply chain from prime energy production to conversion to mechanical energy and into fluid power will have losses. A typical example is shown, for pumps and motors working near to their best efficiency points, in Figure A6.1.

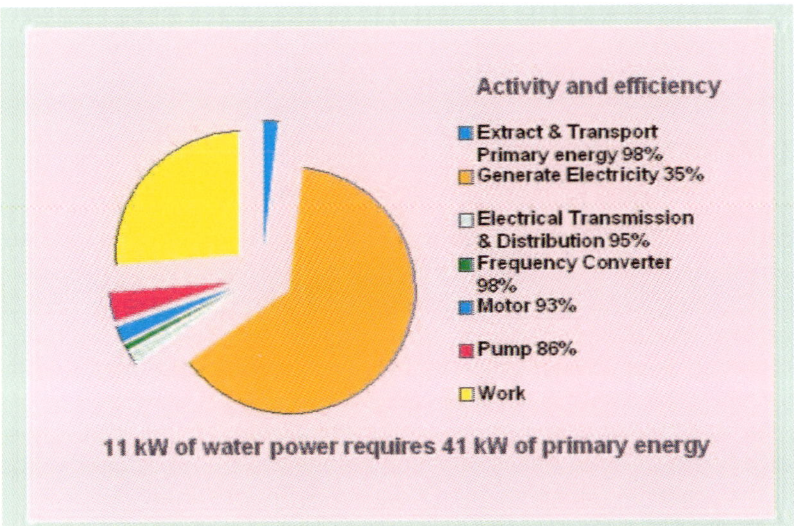

Figure A6.1: Overall illustration of energy efficiency

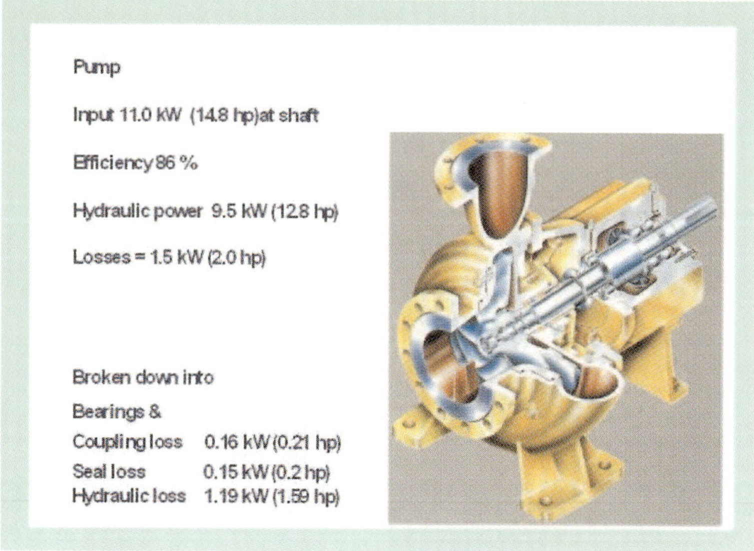

Figure A6.2: Typical rotodynamic pump losses

A6.2 Losses in the pump

The pump industry uses both overall pump efficiency and hydraulic efficiency. Hydraulic efficiency excludes mechanical losses in the pump i.e. bearings, coupling and seal losses (see Figure A6.2).

The losses are shown at a single duty point, and whilst the bearing and seal losses will vary with changes in speed and load, the major effect of speed change is on the hydraulic losses, as shown in Chapter 4.

For a rotodynamic pump running with high friction losses, the hydraulic efficiency may well remain constant although the power absorbed varies with the speed cubed for a friction loss system curve.

A6.3 Motor losses

Typical motor losses are shown in Figure A6.3. A frequency converter drive will cause additional losses in a motor because of the voltage harmonics present in the inverter output. The magnitude of these losses will vary with the motor design. As the resistance to reactance ratio of small motors is higher than that of larger motors, the harmonic components of the current may produce a relatively higher heat loss in a small motor than in a larger one.

In the majority of cases these additional losses will be relatively small with a modern inverter and motor combination, and the additional losses within the motor at full speed and power will be less than 5% of the total motor losses.

In a cage induction motor, the iron losses stay constant with constant

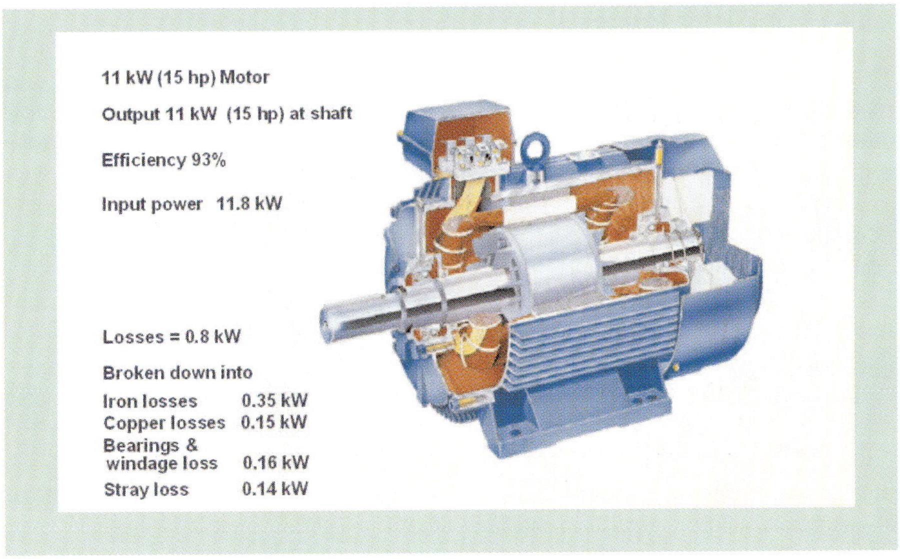

Figure A6.3: Typical motor losses

flux (conventional scalar control) but can reduce with more sophisticated flux control.

Copper losses will reduce with load, while the bearing and windage losses reduce with speed. The distribution of losses in a motor is specific to a particular type of motor.

Figure A6.4 shows the typical efficiency/speed curves for a VFD driven high efficiency type motor with a nominal efficiency of 95% at different percentage torque loads (T/Tn) and speeds with a typical rotodynamic pump speed/torque curve superimposed.

A positive displacement pump with its constant speed torque relationship will typically follow the 100% T/Tn line in Figure A6.4.

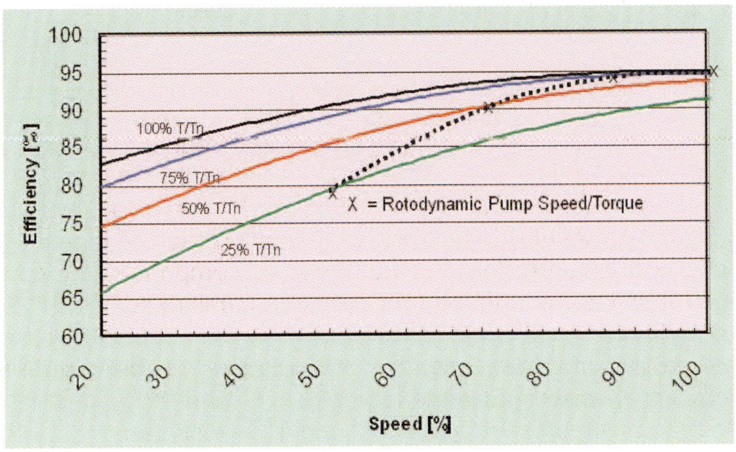

Figure A6.4: Motor efficiency v speed (nominal 95% efficient motor)

A6.4 Converter losses

The converter losses consist of no load and load losses. No load losses mainly consist of the switching losses of the semiconductors, losses in the discharge and snubber resistors, leakage in the capacitors plus the power absorbed by auxiliary supply circuits and fans.

Load losses are generally related to the current passing in the particular components, and the associated main circuit chokes and smoothing components.

Even when the drive is powering a motor at no load, a substantial magnetizing current will be flowing in the inverter Chapter, and therefore the losses are higher than the no load losses.

The distribution of losses within a variable frequency drive is typically as shown in Figure A6.5. Load losses vary widely with the current drawn, because they are proportional to current raised to the power of I.5. Indeed the inverter losses will be increased disproportionately with motors of lower power factor, and thus higher current.

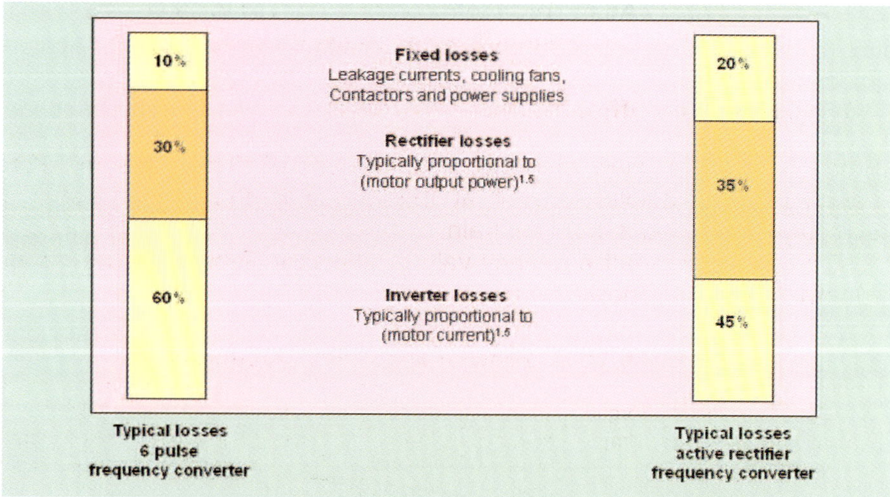

Figure A6.5: Typical losses for frequency converters

Figure A6.6 graphically illustrates the variation of losses with output frequency and load. In order to achieve these high efficiency numbers it is imperative that drives and motors are carefully matched. In order to achieve optimum drive train efficiency it is recommended that high efficiency motors are used, because low efficiency motors suffer higher additional losses when connected to a VFD.

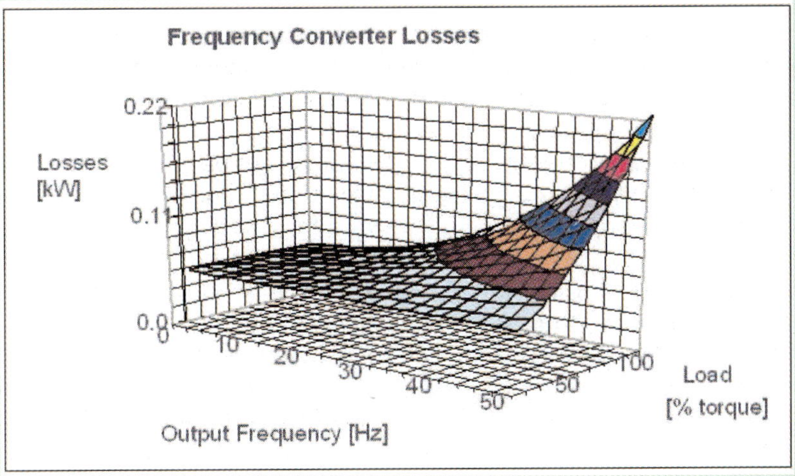

Figure A6.6: Typical losses for an 11 kW VFD up to 50Hz base speed

A6.5 Drive train efficiencies

From the point of view of considering system efficiency, it is therefore generally more helpful to look at the losses in the various items of equipment throughout the drive train.

Typical wire to shaft efficiency values, shown in Table A6.1 are for the rotodynamic pump shown in Figure A6.2.

Table A6.1: Drive train efficiency calculations

Speed [r/min]	Frequency [Hz]	Torque [Nm]	Total power input [kW]	Motor Power out put [kW] (hp)	Converter loss [kW]	Motor loss [kW]	Drive train total efficiency [%]
1440	49.9	73.0	12.17	11.0 (15)	0.27	0.9	90.3
1330	45.8	61.0	9.55	8.5 (11.4)	0.23	0.82	89.0
1210	41.7	51.3	7.48	6.5 (8.7)	0.19	0.79	86.8
1120	37.6	40.9	5.63	4.8 (6.4)	0.17	0.66	85.2
1000	33.6	32.5	4.09	3.4 (4.6)	0.15	0.54	83.1

It can therefore be seen that prediction of efficiencies produce seemingly low numbers, due mainly to no load losses, and the drive manufacturers are often reticent to publish these, however, when these are associated with the savings attributable to the load, the overall effect is generally a substantial saving.

A6.6 Measuring efficiency

Various calorimetric methods have been proposed by research institutes to accurately measure individual component (motor and drive) losses, however, these are impractical for industrial use.

The optimum method of measuring the string (wire to water or overall) efficiency of a pumping system is on a string test in the manufacturer's works. Experience has however shown that this is also very difficult to conduct, due to the disturbing influences of the drive harmonics on the factory network, especially with relatively large powers.

In all cases where these measurements are being made, it is suggested that a power quality analyser is often the most suitable device for recording the results. The values of fundamental current and voltage are generally repeatable from site to site, whereas harmonic currents and the resultant harmonic voltages are a site specific phenomenon.

Even where sophisticated harmonic attenuation measures are fitted, it is possible to see spurious values due to the standing distortion on the network.

Appendix A7: Non VFD drives

A7.1 Summary

This Guide has concentrated on electrical types of variable speed drives, particularly variable frequency drives, because of their many benefits, but Table A7.1 describes briefly the other types of speed changing devices and also variable speed prime movers, which are sometimes used for pumps.

Table A7.1: Other types of variable speed drives

Drive Type	Power Range KW (hp)	Speeds		Drive Efficiency		Advantages	Limitations
		Max r/min	Max Ratio	At Max Speed	At Half Speed		
Variable speed pulleys	Up to 125 Up to (170)	Up to 4500 at low power	4:1	85-90	85-90	Low cost.	Limited power range. Automatic control difficult. Belt life limited.
Eddy current coupling	Up to 1500 (Up to 2000)	Less than input	10:1	96	45-50	Reliability good. Automatic control easy.	Low efficiency at reduced speed. Water cooling required for high powers. Only speed reduction possible.
Fluid coupling	50 to 8000 (70 to 11 000)	Less than input	4:1	97	45-50	Reliability good. Available to high powers at which they become more cost effective.	Low efficiency at reduced speed. Heat exchanger required for high powers. Only speed reduction possible. Automatic control expensive.

Table A7.1: Other types of variable speed drives continued

Drive Type	Power Range KW (hp)	Speeds		Drive Efficiency		Advantages	Limitations
		Max r/min	Max Ratio	At Max Speed	At Half Speed		
Geared Fluid Coupling	300-50 000 (400-70 000)	10,000	4:1	95	43-48	Reliability good. Available to high powers at which they become more cost effective. Speed increase (Gearbox side) and reduction (Fluid Coupling side) possible. Lubrication system can supply driven unit bearings.	Low efficiency at reduced speed. Heat exchanger & lubrication system required for high powers. Automatic control possible but expensive.
Mechan-ical Variators	Up to 120 (Up to 160)	5000 at low powers	12:1	85-95	85-95	Speed increase possible. Good for high Viscosity fluids. Popular for PD Pumps.	Limited power range. Limited life. Careful lubrication and maintenance essential. High torque at low speed.
Devices to vary the supply voltage at AC-motors	Up to 5.5 kW	Less than synchronous	1.5:1	95	43-48	Low cost easy automatic control.	Low speed ratio, low efficiency at reduced speed, most common using only at wet rotor motors, low torque at reduced speed.

Table A7.1: Other types of variable speed drives continued

Drive Type	Power Range KW (hp)	Speeds		Drive Efficiency		Advantages	Limitations
		Max r/min	Max Ratio	At Max Speed	At Half Speed		
IC Engine	1 to 10 000 (1.5 to 15 000)	4000 For Small sizes	4:1	30*	25*	Independent of electrical supply	Inefficient. Requires regular maintenance Requires additional cooling. High Noise level. Engine exhaust.
Steam Turbine	100 to 30 000 (150 to 45 000) and higher	6000	3:1	45*	30*	Independent of electrical supply. Very large sizes High speed.	Needs steam raising plant. Expensive installation.
Gas Turbine	100 to 30 000 (150 to 45 000) and higher	15 000	2:1	40*	30*	Independent of electrical supply. Very large sizes.	Requires a gear box to reduce speed Requires regular maintenance. May require auxiliary services.

* Efficiency values are very size dependent and include the conversion of fossil fuel, or steam, to mechanical power

Appendix A8: Glossary

A8.1 Abbreviations of terms

Term or abbreviation	Description
AC	Alternating current (voltage)
Bar	One atmosphere (14.7 psig)
BEP	Best efficiency point
CE Mark	Indication of compliance with EU Directives
CEMEP	European Committee of Manufacturers of Electrical Machines and Power Electronics
CENELEC	European Committee for the Electro technical Standardisation – responsible for the preparation of harmonised (EN) standards
Choke	Electrical inductance
CIM	Cage induction motor – induction motor with squirrel cage rotor.
CISPR	Commité International Spécial des Perturbations Radioélectriques. (English – International Special Committee on Radio Interference). Prefix for standards prepared by this organisation (Published by the IEC q.v.)
Closed valve head	Head at zero flow
Cogging	Erratic movement of rotor
C/min (CPM pg82)	Cycles per minute
CSI	Current source inverter

Term or abbreviation	Description
D	Diameter (of impeller)
DC	Direct current (voltage)
DCS	Distributed control system
Dead head	Zero flow condition in a PD pump
DOL	Direct-on-line
Drive	Converter
DTC	Direct torque control
DU/dt Filter	Rate of rise filter
$	Dollar
E_s	Specific energy
Earth/Earthing	Ground/Grounding – terms used interchangeably
EMC	Electromagnetic compatibility
EMI	Electromagnetic interference
EN	EuroNorm – Standard issued by CEN or CENELEC, normally prefixed by the national issuing body e.g. BS, DIN etc.
EPAct	Energy Policy and Consultation Act
EU	European Union
€	Euro
F	Frequency
g	Gravitational constant
GAMBICA	Association for Instrumentation, Control and Automation – drive manufacturers trade association (UK)
GTO	Gate turn off
H	Head
HEM	High efficiency motor
h p.	Horse power
HVAC	Heating, ventilating and air conditioning
IEC	International Electro technical Commission – International Standardisation Body Prefix for standards prepared by this organisation
IEEE 519	American Standard – IEEE Recommended Practices and Requirements for Harmonic Control in Electrical Power Systems

Term or abbreviation	Description
IGBT	Insulated gate bi-polar power transistor
Impedance	Electrical term for restriction to flow of current, due to resistors, capacitors and inductors.
Inductor	Electrical inductance
k	Constant value
kWh	Kilowatt hour
LCC	Life cycle cost
LCI	Load commutated inverter
MTBF	Mean time between failures
N	Rotational Speed
NEMA	National Electrical Manufacturers Association (of the US)
NDE	Non drive end (motor bearing)
"noise"	Electromagnetic Interference
NPIP	Net positive inlet pressure
NPSH	Net positive suction head
NPSHA	Net positive suction head available
NPSHR	Net positive suction head required
OD	Outside diameter (of impeller)
P	Power
p	Density
P_{in}	Power input
P_{out}	Power output
PAM	Pulse amplitude modulated
PD	Positive displacement
PDS	Power drive system (converter & motor)
PWM	Pulse width modulated
PLC	Programmable logic controller
PM	Permanent magnet
PTC	Positive temperature coefficient
η	Efficiency
Q	Flow

Term or abbreviation	Description
r.m.s.	Root mean square
Reactor	Electrical inductance
REMA	Rotating electrical machines association (UK)
r/min	Revolutions per minute
RTD	Resistance temperature detector
SCIM	Squirrel cage induction motor
Shut off head	Head at zero flow
SRD	Switched reluctance drive
String Test	Test of complete package, pump, motor, drive etc. Usually carried out at the manufacturers' test facility
TEFC	Totally enclosed fan cooled
TEFV	Totally enclosed fan ventilated
THD	Total harmonic distortion
Transmission	Couplings, gears, belt drive, etc
U	Voltage
U/F	Voltage/frequency relationship
UL	Underwriters Laboratories – American inspection and certification organisation.
V	Volts
VE	Volumetric Efficiency
VSD	Variable speed drive
VSI	Voltage source inverter
VFD	Variable frequency drive
VVVF	Variable voltage variable frequency (drive)
Wire to Water efficiency	Overall efficiency of drive, motor and pump